Annals of Mathematics Studies

Number 125

Combinatorics
of
Train Tracks

by

R. C. Penner

with

J. L. Harer

PRINCETON UNIVERSITY PRESS

PRINCETON, NEW JERSEY
1992

The Annals of Mathematics Studies are edited by
Luis A. Caffarelli, John N. Mather, John Milnor, and Elias M. Stein

Princeton University Press books are printed on acid-free
paper, and meet the guidelines for permanence and durabil-
ity of the Committee on Production Guidelines for Book
Longevity of the Council on Library Resources

Printed in the United States of America
by Princeton University Press, 41 William Street
Princeton, New Jersey

Library of Congress Cataloging-in-Publication Data

Penner, R. C. 1956–
    Combinatorics of train tracks / by R. C. Penner with J. L. Harer.
    p.   cm. — (Annals of mathematics studies ; no. 125)
    Includes bibliographical references.
    ISBN 0-691-08764-4 (cloth)
    ISBN 0-691-02531-2 (pbk.)
    1. Geodesics (Mathematics)   2. CW complexes.   3. Combinatorial
analysis.   I. Harer, J. (John), 1952- .   II. Title.   III. Title: Train tracks.
IV. Series.
QA649.P38   1991
511'.6—dc20   91-33559

Dedicated to my Father,
Dr. Stanford S. Penner

CONTENTS

PREFACE

We study here one aspect of the mathematics pioneered by William P. Thurston, namely, the rich combinatorial structure of the space of measured geodesic laminations in a fixed surface. One may think of this space as a natural completion of the collection of all $\mathbb{R}_+$-weighted isotopy classes of essential simple closed curves in the surface, and it turns out to be a piecewise-linear manifold homeomorphic to a Euclidean space whose dimension depends upon the topological type of the surface. Roughly, a train track is a CW complex in the surface (together with extra structure), and appropriate train tracks correspond to charts on this manifold. More explicitly, a measure on a train track is an assignment of a nonnegative real number (satisfying certain conditions) to each edge of the underlying CW complex, and a measured train track gives rise to a corresponding measured geodesic lamination in the surface. We may thus explore the space of measured geodesic laminations by studying measured train tracks in the surface.

Our techniques are almost always essentially combinatorial. For instance, the first chapter (which includes the basic Thurston theory of train tracks) involves numerous analyses of separation properties of points in the circle at infinity of the hyperbolic plane. A combinatorially defined equivalence relation on measured train tracks (implicit in [T2] and considered in [P1]) is systematically studied in the second chapter. Our investigation of this equivalence relation leads to a family of standard models (from [P1]) for the equivalence classes. In the third chapter, we rely on these standard models to study both the local and global structure of the space of measured geodesic laminations and derive many basic results of Thurston. Among these are certain global coordinates (from [P1]), called the Dehn-Thurston coordinates, on the space of measured geodesic laminations. There is thus much from [P1] which finds application in our approach.

Because of the combinatorial nature of our arguments, there are few real pre-requisites to this volume. We take for granted basic differential topology (at the level, say, of [GP]), certain standard facts about surface topology (see [E]), and at least a working knowledge of elementary hyperbolic geometry at the level of [T2;§2]. (See [Be] for a more extensive treatment.) Furthermore, we assume certain elementary facts about Fuchsian groups, the treatment in [A] being much more than sufficient. Fundamental to us are two results: the Uniformization Theorem of Köebe, Klein, Poincaré, and Ahlfors-Bers (which we use simply to guarantee that the surfaces we consider have the hyperbolic plane as universal cover), and the Nielsen Extension Theorem (which says that a homeomorphism of such a surface lifts and extends to a homeomorphism of the circle at infinity). A suitable reference for both of these basic results is [A]. The reader willing to take these results for granted should be able to tackle this volume with no further pre-requisites. Our discussion is completely independent (except for the Epilogue) of the parallel theory of measured foliations, but of course the reader familiar with Thuston's work from [FLP] will greatly profit in insight.

The action of homeomorphisms of a surface on curves in the surface extends to a continuous action of the mapping class group on the space of measured geodesic laminations. Train tracks are quite useful for studying the dynamics both of this action and of a given homeomorphism on the surface itself. By and large, we treat only the "static" theory of measured geodesic laminations, in that we only rarely consider this action of the mapping class group. The two exceptions to this are in the Epilogue, where we survey aspects of the larger contexts of Riemann surfaces and surface automorphisms to which our investigations here are relevant, and the Addendum, which includes explicit formulas from [P1] for the action of the mapping class group on Dehn-Thurston coordinates.

In the bulk of the text, we treat surfaces with no boundary ("of the first kind"), relegating the extension of results to the bounded case to remarks and exercises. For bounded surfaces, we discuss in the first chapter a generalization (from [P1]) of train tracks (called train tracks with stops), and the extension of our results to this setting is similarly left to remarks and exercises. The rationale for this is that these extensions are often straight-forward while the particulars of them may obfuscate the relevant material.

ACKNOWLEDGEMENTS

First and foremost, thanks go to Bill Thurston for being generous with his time during 1982-83 and 1984-85, when much of this work was done. Early drafts of certain sections (indeed most of §1.1, 1.3-1.5 and 3.2) represent joint work with John Harer, with whom there were many invigorating discussions. The proof of Theorem 1.4.4 was also kindly contributed by Nat Kuhn. There are intellectual debts to the lecture notes [Cb] and substantial such debts to [Tg], upon which §1.1 and 1.3 are roughly based. We thank Scott Wolpert for helpful comments on an early version of this manuscript and Max Bauer for clarifications of various arguments in the first chapter. Matt Grayson, Steve Kerckhoff, and especially Lee Mosher made many valuable comments.

It is a pleasure to finally acknowledge the support of the National Science Foundation as well as the warm hospitality of Insitut Mittag-Leffler during 1983-84 and Stanford University, especially Jim Milgram, during Spring, 1991.

COMBINATORICS OF TRAIN TRACKS

CHAPTER 1   THE BASIC THEORY

We begin with the basic definitions and ideas. Much of the material of this chapter is due to Thurston. Our treatment includes new results as well as new proofs of facts from [T1], [T2] and [Tg]. We define train tracks and introduce the notion of a trainpath on a train track, motivate the notion of transverse measure on a train track by considering a sense in which train tracks carry curves, and recall Dehn's parametrization of curves in surfaces. The notions of recurrence, transverse recurrence, and tangential measure are introduced, and we provide several equivalent definitions. Generic train tracks (to which we restrict attention in later chapters) are introduced, and a certain geometric condition is shown to be equivalent to transverse recurrence in this setting. The importance of transverse recurrence as a technical condition is then highlighted by several results on trainpaths in transversely recurrent tracks. Laminations are defined, their elementary properties are explored, and an explicit construction of a measured geodesic lamination from a measured transversely recurrent train track is given. Surfaces with boundary are then considered, and the foregoing theory is seen to apply with essentially no modifications; we also introduce a relative notion of train tracks, called train tracks with stops (which arose in [P1]), and explore the basic theory in this setting.

§1.1 TRAIN TRACKS

Let $\hat{F}_g^s$ be a closed, smooth, oriented surface of genus g with $s \geq 0$ distinguished points, whose union we denote Δ. We will usually regard Δ as deleted from the surface and define $F_g^s = \hat{F}_g^s - \Delta$ so that each point of Δ gives rise to a cusp (or puncture) of $\hat{F}_g^s$. When the topological type of the surface F_g^s is fixed or not important, we may call it simply F. Suppose that $\tau \subset F$ is a finite collection of one-dimensional CW complexes, each made up of vertices, called *switches*, and edges, called *branches*, disjointly imbedded in F. Branches are open one-cells by convention so that τ is the disjoint union of its switches and branches. If b is a branch of τ and $p \in b$, then by a *half-branch* of b, we mean a component of $b - \{p\}$. Two half-branches of a given branch b are regarded as equivalent if their intersection is again a half-branch of b, and the equivalence class is called an *end* of b. Thus, there are two ends of each branch of τ. An end e of a branch b is said to be *incident* on a switch v of τ if v lies in the closure of a half-branch representing e. The *valence* of a switch of τ is the number of distinct ends which are incident on it.

We say that $\tau \subset F$ is a *train track* (or simply a *track*) in F provided that the following conditions hold.

(1) (Smoothness) τ is C^1 away from its switches. For each switch v of τ, there is furthermore a line $T_v(\tau)$, called the *tangent line* to τ at v in the tangent plane to F at v, so that the following condition holds for each half-branch $\hat{b}$ whose closure contains v: a C^1 parametrization $\beta : (0,1) \to \hat{b}$ with $\lim_{t \to 1} \beta(t) = v$ extends to a C^1 map on $(0,1]$ so that the one-sided derivative at 1 lies in $T_v(\tau)$. We may take the parametrization β in such a way that $w = \lim_{t \to 1} \beta'(t) \neq 0$. The unit vector in the direction of w lies in $T_v(\tau)$, is independent of the choices above and is called the *one-sided tangent vector* to the end e corresponding to $\hat{b}$.

(2) (Non-degeneracy) For any switch v of τ, there is an imbedding $f : (0,1) \to \tau$ with $f(\frac{1}{2}) = v$ which is a C^1 map into F. Thus, no

4

switch of τ is univalent, and we moreover demand that each simple closed curve component of τ contains a unique bivalent switch and that every other switch of τ is at least trivalent.

(3) (Geometry) Suppose that S is a component of $F - \tau$, and let $D(S)$ denote the double of S along the C^1 frontier edges of S. Thus, non-smooth points in the frontier of S give rise to punctures of $D(S)$. We require that the Euler characteristic $\chi(D(S))$ of $D(S)$ be negative.

Examples of train tracks may be found in Figure 1.1.1.

FIGURE 1.1.1

Condition (1) requires that $\tau \subset F$ be a "branched one-submanifold" of F rather than just a CW complex imbedded in F. Condition (2) rules out "dead ends" along τ and prohibits unnecessary switches while introducing the convenient convention of requiring one bivalent switch on each curve component of τ. We will comment further on condition (3) below.

Suppose that D is a closed disk imbedded in the plane whose boundary is piecewise C^1 with $n \geq 0$ discontinuities in the tangent. If $i: D \to F$ is a C^1 immersion which is an imbedding of the interior $\overset{\circ}{D}$ of D into F, then the image $i(\overset{\circ}{D})$ is an *imbedded n-gon* in F. In case i is only an immersion on $\overset{\circ}{D}$, then $i(\overset{\circ}{D})$ is an *immersed n-gon* in F. In either case, $i(D-\overset{\circ}{D})$ is called the *frontier* of the n-gon. The image under i of a non-smooth point in the frontier of D is called a *vertex* of the n-gon, so the vertices canonically decompose the frontier of an n-gon into n smooth arcs, which are called *frontier edges* of the n-gon. Similarly, if $X \subset \overset{\circ}{D}$ is a finite set of cardinality m and $i: \overset{\circ}{D}-X \to F$ is as above, we analogously define the image $i(\overset{\circ}{D}-X)$ to be an *immersed* or *imbedded m-times-punctured n-gon* in F. We also define an *immersed* or *imbedded n-gon-minus-a-disk* to be the image $i(\overset{\circ}{D}-X)$ $\subset F$ where $X \subset \overset{\circ}{D}$ is an open disk whose frontier is C^1 and lies in $\overset{\circ}{D}$; in particular, a nullgon-minus-a-disk is called an *annulus* in F.

Condition (3) in the definition of train track is equivalent to the condition that no component of $F - \tau$ is an imbedded nullgon, monogon, bigon, once-punctured nullgon, or annulus. In particular, a train track can have no curve component homotopic into a puncture, and, furthermore, no two (distinct) curve components of a train track can be homotopic.

If σ and τ are train tracks in F and σ is contained in τ as a point set, then we say that σ is a *subtrack* of τ and write $\sigma \subset \tau$. In this case, σ can be obtained from τ by deleting certain branches, amalgamating any two (not necessarily distinct) branches which meet at a resulting bivalent vertex, and then inserting the bivalent switch on each curve component of σ.

Define a *(finite) trainpath* on τ to be any C^1 immersion $\rho : [n, m] \to F$, where $n < m$ and $n, m \in \mathbb{Z}$, with image contained in τ so that the restriction of ρ to each subinterval $(k, k + 1) \subset [n, m]$ with $k \in \mathbb{Z}$, is a branch of τ with $\rho(k)$ and $\rho(k + 1)$ switches of τ. The integer $m - n$ is called the *length* of ρ, and ρ is said to be *closed* provided $v = \rho(m) = \rho(n)$ and the tangents $\rho'(m)$ and $\rho'(n)$ (both of which lie in the tangent line to τ at v) are non-zero and point in the same direction. Analogously, in case $n = -\infty$, $m = +\infty$ or both, we define *semi-infinite* and *bi-infinite* *trainpaths*, respectively.

Two trainpaths $\rho_1 : I_1 \to \tau$ and $\rho_2 : I_2 \to \tau$ are regarded as equivalent if there is an orientation-preserving homeomorphism $\phi : \mathbb{R} \to \mathbb{R}$ (or perhaps the restriction of such a map) so that $\rho_1 = \rho_2 \circ \phi$, and when no ambiguities could develop, we may also simply identify a trainpath with the image of an underlying function. The *reverse* of the trainpath $t \mapsto \rho(t)$ is the trainpath $t' \mapsto \rho(-t')$, where the parameter t' is restricted to lie in a range so that this definition makes sense. If ρ is a trainpath in the train track $\tau \subset F$, $\tilde{F}$ is some cover of F, and $\tilde{\tau}$ is the full pre-image of τ in $\tilde{F}$, then we will also refer to a lift of ρ to $\tilde{F}$ as a trainpath on $\tilde{\tau}$. Notice that if $\sigma \subset \tau$, then a trainpath on σ gives rise to a trainpath on τ in the natural way.

Proposition 1.1.1: *Suppose that R is a surface whose boundary ∂R (if any) is piecewise C^1, τ is a train track in the surface F, and $i : R \to F$ is a C^1 immersion on $\overset{\circ}{R}$ so that the restriction of i to each smooth curve or arc in ∂R is a finite trainpath on τ. Then $\chi(D(R)) < 0$, where $D(R)$ is the double of R along the C^1 arcs and curves in ∂R.*

Proof: We can easily add branches to τ to produce a train track $\hat{\tau} \supset \tau$ in F so that each component of $F - \hat{\tau}$ is either a trigon or a once-punctured monogon (where the number of components of the latter type is s), and it clearly suffices to prove the proposition for $\hat{\tau}$. Construct a foliation of each complementary trigon T with one three-pronged singular point $V \in T$ so that the leaves of the foliation are transverse to the frontier edges of T as in Figure 1.1.2a; similarly, foliate each complementary punctured monogon M with one three-pronged singular point $U \in M$ as in Figure 1.1.2b. These foliations combine in the natural way to produce (after some smoothing) a foliation $\mathcal{F}$ of F (which is C^1 away from singular points such as U, V above) so that $\mathcal{F}$ is transverse to $\hat{\tau}$.

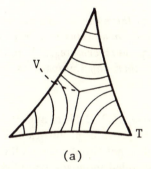
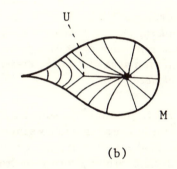

(a)   (b)

FIGURE 1.1.2

The pull-back $i^*(\mathcal{F})$ of $\mathcal{F}$ to R is a foliation of R which has only three-pronged singularities in $\overset{\circ}{R}$; suppose that there are $p > 0$ such singularities and $q \geq 0$ discontinuities on ∂R. Consider the double cover $\tilde{R}$ of R branched over the singularities in $\overset{\circ}{R}$; of course, by the Riemann-Hurwitz formula,

$$\chi(\tilde{R}) = 2\chi(R) - p.$$

The foliation $i^*(\mathcal{F})$ lifts to a foliation of $\tilde{R}$ which is C^1 on the interior except for p six-pronged singularities; furthermore, there are $2q$ discontinuities on $\partial\tilde{R}$. Let us double $\tilde{R}$ along each of its boundary components (in the usual sense of doubling) to produce a surface $\check{R}$. The foliation of $\tilde{R}$, in turn, gives rise to a foliation of $\check{R}$, and we can perturb the latter to finally produce a foliation $\check{\mathcal{F}}$ of $\check{R}$ which is C^1 except for $2p$ six-pronged singularities, say $\{V_j\}$, and $2q$ centers, say $\{U_i\}$, one center arising from each non-smooth point of $\partial\tilde{R}$. Since the number of singularities is even, it is easy to produce a vectorfield on $\check{R}$ which is tangent to $\check{\mathcal{F}}$ with zero set $\{U_i\} \cup \{V_j\}$ so that each U_i has order one and each V_j has order -2. By the Poincaré-Hopf Theorem

$$\chi(\check{R}) = 2q - 4p,$$

and so

$$\chi(\tilde{R}) = \frac{1}{2}\chi(\check{R}) = q - 2p.$$

Finally, since

$$\chi(D(R)) = 2\chi(R) - q,$$

we find

$$\chi(D(R)) = \chi(\tilde{R}) + p - q = q - 2p + p - q = -p < 0,$$

as was claimed.   q.e.d.

Corollary 1.1.2: *If $\tau \subset F$ is a train track, then there can be no immersed nullgon, monogon, bigon, once-punctured nullgon, or annulus in F each of whose frontier edges is a finite trainpath in τ. Furthermore, if $\tilde{\tau}$ is the full pre-image of τ in the universal cover of F, then any trainpath on $\tilde{\tau}$ is imbedded.*

Proof: For any of the surfaces R mentioned above, $\chi(D(R))$ is nonnegative, so the first assertion follows immediately from the previous proposition. For the final assertion, if ρ is a trainpath on $\tilde{\tau}$ which is not simple, then a finite sub-trainpath of ρ forms the frontier of an imbedded monogon or nullgon in the universal cover; this region projects to an immersed monogon or nullgon in F, and this contradicts the result just proved.   q.e.d.

Corollary 1.1.3: *A surface $F = F_g^s$ contains a train track if and only if $\chi(F) < 0$ (i.e., $2g + s - 2 > 0$) and F is not the surface F_0^3 (i.e., if $g = 0$, then $s > 3$). Furthermore, if $\tau \subset F$ is a train track with v switches and b branches, then we have the inequalities*

$$v \leq -6\chi(F) - 2s,$$

$$b \leq -9\chi(F) - 3s.$$

Proof: That the stated conditions are necessary for F to contain a train track follows from the previous proposition (where $\partial F = \emptyset$, so $\chi(D(F)) = 2\chi(F)$) and the observation that F_0^3 contains no train track (since every essential curve in this surface is puncture-parallel). Sufficiency is easily established by producing an essential non-puncture-parallel curve in F (and adding a bivalent vertex) to produce a train track in F.

For the second part, we may complete τ to a train track $\hat{\tau}$ as in the proof of Proposition 1.1.1 and then alter $\hat{\tau}$ slightly near its switches as in Figure 1.1.3 to produce a train track $\check{\tau}$ each of whose switches is trivalent.

If $\check{\tau}$ has $\check{v}$ switches and $\check{b}$ branches, then $\check{v} \geq v$ and $\check{b} \geq b$, and we let $\check{t}$ denote the number of trigons complementary to $\check{\tau}$ in F. Since each switch of $\check{\tau}$ is trivalent,

$$2\check{b} = 3\check{v},$$

and since there are s complementary punctured monogons,

$$3\check{t} + s = \check{v}.$$

Meanwhile, it was proved above that

$$\check{t} + s = -\chi(D(F)) = -2\chi(F),$$

so the first inequality follows from

$$v \leq \check{v} = 3\check{t} + s = -6\chi(F) - 2s$$

and the second from

$$b \leq \check{b} = \frac{3}{2}\check{v} \leq -9\chi(F) - 3s.$$

q.e.d.

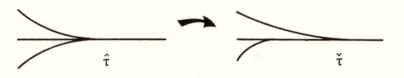

FIGURE 1.1.3

In light of the previous result, we will henceforth tacitly assume that any surface $F = F_g^s$ considered has negative Euler characteristic and is not the surface F_0^3.

§1.2 MULTIPLE CURVES AND DEHN'S THEOREM

A primary motivation for studying train tracks is that large numbers of curves may be described by using a single track. A family C of smooth simple closed curves disjointly imbedded in $F = F_g^s$ so that no component of C is either null-homotopic or homotopic into a puncture (to be referred to as *puncture-parallel*) is called a *multiple curve* in F. If C and D are each multiple curves, then we define the *geometric intersection number* of C and D, denoted $\iota\,(C, D)$, to be the minimum of $\mathrm{card}(\gamma \cap \delta)$ over all multiple curves γ and δ which are isotopic to C and D, respectively, where $\mathrm{card}(X)$ denotes the cardinality of the finite set X.

The multiple curve C is *carried* by the train track τ if there is a C^1 map $\phi\colon F \to F$, called the *supporting map*, such that

(1) $\phi(C) \subset \tau$,

(2) ϕ is homotopic to the identity map $F \to F$, and

(3) The restriction of the differential $d\phi_p$ to the tangent line to C at p is nonzero for every $p \in C$.

One thinks of ϕ as squashing together strands of C which are "nearly" parallel. We also say that the train track σ is *carried* by the train track τ using the same definition; Condition (3) above is legitimate by Condition (1) in the definition of train track. Write $\sigma < \tau$ or $\tau > \sigma$ if τ carries σ, and notice that $<$ is a transitive relation on train tracks. Furthermore, if $\sigma < \tau$, then a closed trainpath on σ gives rise to a closed trainpath on τ in the natural way.

Multiple curves carried by τ may be described by labeling the branches of τ with integers: for each branch b, pick $p \in b$ and write $\mu_C(b)$ for the number of points in $\phi^{-1}(p) \cap C$, where ϕ is the supporting map of the carrying $C < \tau$. The definition is independent of the choice of $p \in b$, and these integers satisfy a condition defined presently. (In fact, μ_C is also independent of the choice of supporting map; see Proposition 1.7.5.)

10

For each switch v of τ, fix a direction in the tangent line $T_v(\tau)$ to τ at v. The end e of a branch b of τ which is incident on v may then be called *incoming* if the direction of the one-sided tangent vector to e at v agrees with this direction, *outgoing* if not. The function μ_C defined above gives rise to a function defined on ends of branches by the rule $\mu_C(e) = \mu_C(b)$ if e is an end of b. If $e_1, \ldots, e_r$ are the incoming ends of branches incident on v and $e_{r+1}, \ldots, e_{r+t}$ are outgoing, then this function satisfies the following *switch condition at v*:

$$\mu_C(e_1) + \ldots + \mu_C(e_r) = \mu_C(e_{r+1}) + \ldots + \mu_C(e_{r+t}).$$

Insofar as μ_C satisfies the switch condition at each vertex of τ, we say simply that μ_C satisfies the *switch conditions (on τ)*.

Conversely, any assignment μ of a nonnegative integer to each branch of τ which satisfies the switch conditions defines a multiple curve as follows. Choose a regular neighborhood N of τ in F and arrange $\mu(b)$ arcs parallel to the branch b disjointly imbedded in N as in Figure 1.2.1a. If v is a switch of τ, the number of endpoints of arcs so constructed corresponding to incoming ends of branches agrees with the number corresponding to outgoing ends by the switch condition at v. There is thus a unique way to combine these arcs in N near v, connecting incoming to outgoing, so that the result is a collection of arcs disjointly imbedded in N; see Figure 1.2.1b. Performing this construction near each switch of τ yields a collection of simple closed curves disjointly imbedded in $N \subset F$. Since there can be no immersed nullgon or punctured nullgon in F whose frontier consists of a finite trainpath on τ by Corollary 1.1.2, it follows that no curve in the collection just constructed is null-homotopic or puncture-parallel; thus, the collection is in fact a multiple curve, as was asserted.

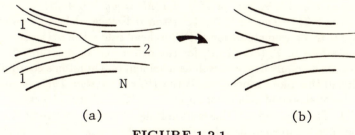

(a)   (b)

FIGURE 1.2.1

We must generalize the material introduced above for later application as follows. Suppose that $C \subset F$ is a collection of closed curves immersed in F; components of C need *not* be simple and may intersect one another in F. If $\tau \subset F$ is a train track, then we say that τ *carries* C if there is a homotopy ϕ_t, for $0 \leq t \leq 1$, with ϕ_0 the given immersion of C in F, so that $\phi = \phi_1$ satisfies conditions (1) and (3) above; ϕ is called the *supporting map* of the carrying. As in the previous paragraph, if C is carried by τ, then no component of C can be null-homotopic or homotopic into a puncture by Corollary 1.1.2. Furthermore, if C is carried by τ, then there is an induced $[\mathbb{Z}_+ \cup \{0\}]$-valued function μ_C defined as before, where the count of points in $\phi^{-1}(p) \cap C$ (for p a regular value of the supporting map ϕ) must be performed with multiplicity. It is clear that such a function μ_C must satisfy the switch condition at each switch of τ, and by the previous paragraph, therefore gives rise to a multiple curve, say $C' \subset F$. In particular, a closed trainpath ρ on τ gives rise to a (connected) curve, say C, immersed in F which is carried by τ, and C in turn gives rise to a multiple curve C' (which need not be connected), as above. Notice that if ρ traverses a branch b of τ exactly m_b times, then $\mu_C(b) = \mu_{C'}(b) = m_b$.

Define a *multicurve* in F to be the isotopy class of a multiple curve in F, and let $\mathcal{S}(F)$ denote the set of all multicurves in F. We say that a multicurve is *connected* if some underlying multiple curve is connected; the collection of connected multicurves in F is denoted $\mathcal{S}'(F) \subset \mathcal{S}(F)$. The geometric intersection number extends to a function

$$\iota: \mathcal{S}(F) \times \mathcal{S}(F) \to \mathbb{Z}_+ \cup \{0\}$$

in the natural way. We say that a train track τ *carries* a multicurve if it carries some representative multiple curve. (There is actually a one-to-one correspondence between the collection of all multicurves carried by a train track and the collection of all $[\mathbb{Z}_+ \cup \{0\}]$-valued functions which satisfy the switch conditions defined on the set of branches of τ; see Theorem 1.7.12.)

In a 1922 Breslau lecture [D], Dehn described a one-to-one correspondence between the set $\mathcal{S}(F_g^s)$ of multicurves in a surface of negative Euler characteristic and a subset of $\mathbb{Z}^M$, for $M = 6g - 6 + 2s$; we call this result *Dehn's Theorem*. In 1976, Thurston rediscovered Dehn's result and extended it to a parametrization of (Whitehead equivalence classes of) measured foliations in F. (See [FLP] for the definitions of measured foliation and Whitehead equivalence as well as a version of Thurston's coordinates.) We will call this more general result the *Dehn-Thurston Theorem* and will derive the analogue of this theorem in the setting of train tracks in Theorem 3.1.1. The remainder of this section is devoted to a discussion of Dehn's Theorem itself; sufficient detail is given that the reader could (and should) supply a complete proof. [FLP] contains a proof of the Dehn-Thurston Theorem in the setting of measured foliations (this theorem follows from our train track version), and [P1] introduced the version of Dehn's Theorem which we discuss here. In the Addendum, we briefly describe the computations of [P1], which give the natural action of the mapping class group of

F (that is, the isotopy classes of orientation-preserving homeomorphisms of F) on Dehn's coordinates for multicurves in F.

A *pair of pants* is a compact planar surface of Euler characteristic -1; thus, a pair of pants is homeomorphic to a closed disk minus two open disks whose closures are disjoint and lie in the interior of the closed disk. Choose an oriented "standard" pair of pants P which is C^1 and has C^1 boundary. Label the boundary components ∂_i and choose closed arcs $w_i \subset \partial_i$, called *windows*, for $i = 1, 2, 3$ as in Figure 1.2.2a.

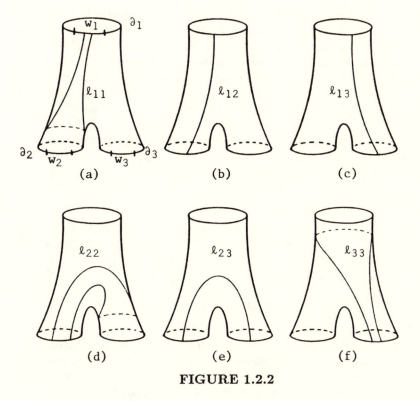

FIGURE 1.2.2

Let D be a collection of arcs properly imbedded in P with ∂D contained in the interior of the windows so that no component of D is boundary-parallel (i.e., properly homotopic into ∂P). We say that two such collections are *parallel* if they are related by a proper isotopy of the identity map $P \to P$ which fixes each point of $\partial P - \cup\{w_1, w_2, w_3\}$. A basic fact (left as an exercise) due to Dehn is that there is a proper isotopy of D (not necessarily respecting the windows) which takes D into a disjoint union

of parallel copies of the arcs illustrated and labeled in Figure 1.2.2. The
different possibilities are uniquely determined by the number m_i of times
that D intersects ∂_i, $i = 1, 2, 3$, subject only to the restriction that the sum
$m_1 + m_2 + m_3$ is even. Moreover, D is parallel to a disjointly imbedded
collection of arcs so that each arc in the collection is parallel to an arc
obtained from one of those illustrated in Figure 1.2.2 by performing Dehn
twists along certain of the components of ∂P. For each triple (m_1, m_2, m_3)
with $m_1 + m_2 + m_3$ even, let us choose a collection of imbedded arcs (with
no twisting and with endpoints in the interiors of the windows), which
represents the corresponding proper isotopy class of one-submanifold of P.

The simple structure of one-manifolds properly imbedded in P suggests
that we decompose an arbitrary surface $F = F_g^s$ (of negative Euler characteristic) into pairs of pants. A *pants decomposition* $\{K_i\}$ of F is a multiple
curve in F so that each component R of $F - \cup\{K_i\}$ is homeomorphic to the
interior of a pair of pants. We do not require the closure of R in F to be an
imbedded pair of pants. Some examples of pants decompositions are given
in Figure 1.2.3. Each component K_i of a pants decomposition is called a
pants curve. By considering the Euler characteristic, one finds that there
are $N = 3g - 3 + s$ pants curves in a pants decomposition of $F = F_g^s$ and
$M = 2g - 2 + s$ complementary regions. In the following discussion, the
subscript i ranges from 1 to N, and the subscript j ranges from 1 to M.

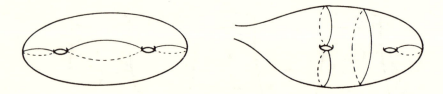

FIGURE 1.2.3

For each pants curve K_i, choose a small closed regular neighborhood
A_i of K_i. Let $A = S^1 \times [-1, 1]$ denote the "standard" oriented annulus and
let G be the canonical projection collapsing A onto the core $S^1 \times 0$. Choose
for each index i an orientation-preserving homeomorphism ν_i of A onto A_i
which maps $S^1 \times 0$ to K_i; ν_i is called the *characteristic map* of the annulus
A_i. In each K_i, choose a closed arc α_i.

Choose an open regular neighborhood $\mathcal{U}$ of the set Δ of punctures of F
so that the frontier of $\mathcal{U}$ consists of smooth curves, one about each puncture.
Each component P_j of $F - \mathcal{U} - \cup\{\overset{\circ}{A}_i\}$ is a pair of pants imbedded in F,
and we may choose for each P_j an orientation-preserving homeomorphism
f_j of P onto P_j which carries each component of $f_j^{-1} \circ \nu_i \circ G^{-1} \circ \nu_i^{-1}(\alpha_i)$ to

a window in ∂P, whenever $A_i \cap P_j \neq \emptyset$. The homeomorphism f_j is called the *characteristic map* of P_j.

Now, let C be a multiple curve in F representing some multicurve, and for each index i, let

$$m_i = \iota(C, K_i),$$

where we assume that C and K_i *hit efficiently* in the sense that there is no imbedded bigon in F whose frontier consists of one arc from each of C and K_i; in this case, one observes that C and K_i themselves realize the geometric intersection number of the corresponding multicurves. (See, for instance [FLP].) The parameter

$$m_i : \mathcal{S}(F) \rightarrow [\mathbb{Z}_+ \cup \{0\}]$$

is called simply an *intersection number*. We may isotope C so that card($C \cap \nu_i(\mathrm{S}^1 \times \{t\})$) $= m_i$, for all i and all $t \in [-1, 1]$ arranging that $C \cap \partial A_i \subset \nu_i \circ G^{-1} \circ \nu_i^{-1}(\alpha_i)$, for all i. (This assertion is left as an exercise.) Next, let us isotope into each annulus A_i any component of C which is itself isotopic to K_i. Finally, using the basic fact about families of arcs properly imbedded in P, we may further isotope C so that $f_j^{-1}(C \cap P_j)$ is one of our chosen models in P, for each index j.

We then define a *twisting number* t_i for each pants curve K_i as follows:

-If $m_i = 0$, take $t_i \geq 0$ to be the number of components of $C \cap A_i$

-If $m_i > 0$, then define $d = G^{-1}(\nu_i^{-1}(*)) \subset A$, where $* \in \partial \alpha_i$, and consider the collection $c = \nu_i^{-1}(C \cap A_i)$ of arcs properly imbedded in A. In this case, $|t_i|$ is defined to be the minimum of card($\gamma \cap d$), where the minimum is over all families γ of arcs properly imbedded in A which are isotopic to c *fixing ∂A pointwise*. The sign of t_i is positive if some component of c twists to the right in the oriented annulus A, and the sign of t_i is negative if some component twists to the left. (Notice that there cannot be components twisting in both directions since C is imbedded, and if there is no twisting, then $t_i = 0$.)

We have indicated how to compute intersection numbers and twisting numbers of a multicurve in F. Dehn's Theorem asserts that these parameters uniquely determine this multicurve.

Theorem 1.2.1 [Dehn's Theorem]: *There is a parametrization of the collection $\mathcal{S}(F_g^s)$ by a subset of $(\mathbb{Z}_+ \cup \{0\})^N \times \mathbb{Z}^N$, where $N = 3g - 3 + s$. The parameter $(m_1, \ldots, m_N) \times (t_1, \ldots, t_N)$ corresponds to a multicurve if and only if the following conditions are satisfied.*

(a) If $m_i = 0$, then $t_i \geq 0$, for each $i = 1, \ldots, N$.

(b)If $K_{i_1}, K_{i_2}, K_{i_3}$ are pants curves which together bound a pair of pants imbedded in F_g^s, then the sum $m_{i_1} + m_{i_2} + m_{i_3}$ of correspond-ing intersection numbers is even. Furthermore, the intersection number corresponding to a pants curve which bounds a torus-minus-a-disk or twice-punctured nullgon imbedded in F_g^s is even.

As we have seen, to parametrize multicurves, we must make a large number of conventions. We define a *basis* $\mathcal{A}$ for multicurves on F to be a choice of pants decomposition of F, choices of associated characteristic maps, and a choice of model arcs on P as in Figure 1.2.2. In practice, to specify a basis, we fix a choice of model arcs in P once and for all, regard F as imbedded in $\mathbb{R}^3$, and draw pictures. The characteristic map ν_i is the trivialization of the normal bundle to $K_i \subset F \subset \mathbb{R}^3$ that extends across a disk in $\mathbb{R}^3$ with boundary K_i. We draw and label K_i and ν_i, and draw $f_j(\ell_{12})$ and $f_j(\ell_{13})$, labeled 2 and 3 respectively, in each P_j.

Example: Consider the basis for $\mathcal{S}(F_2^0)$ indicated in Figure 1.2.4a. We will draw a representative multiple curve C of the multicurve with Dehn parameter values $(m_1, m_2, m_3) \times (t_1, t_2, t_3) = (5, 1, 2) \times (1, -1, 0)$. There are five components of $C \cap A_1$ since $m_1 = 5$, and one of these twists to the right since $t_1 = +1$. Similarly, there is one component of $C \cap A_2$ since $m_2 = +1$, and it twists once to the left since $t_2 = -1$; there are two components of $C \cap A_3$ since $m_3 = 2$ and no twisting since $t_3 = 0$. Thus, we draw our representative C in each of the annuli $A_i, i = 1, 2, 3$, as in Figure 1.2.4b. We then connect up these arcs uniquely using the images under f_1 of arcs parallel to $\ell_{13}, \ell_{23}, \ell_{33}$ and under f_2 of arcs parallel to $\ell_{12}, \ell_{23}, \ell_{22}$ as shown in Figure 1.2.4c.

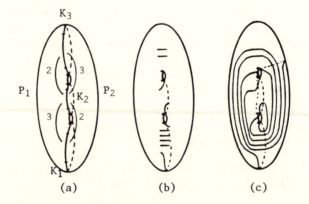

(a)   (b)   (c)

FIGURE 1.2.4

For later use, we define a *standard basis* $\mathcal{A}_g^s$ on F_g^s in Figure 1.2.5 (using the arcs in Figure 1.2.2).

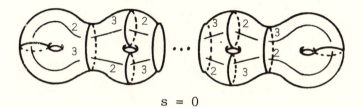

$$s = 0$$

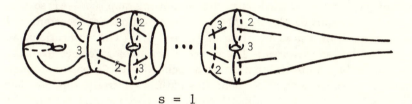

$$s = 1$$

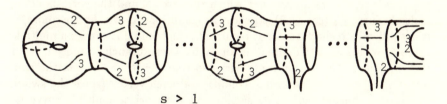

$$s > 1$$

FIGURE 1.2.5

§1.3 RECURRENCE AND TRANSVERSE RECURRENCE

The train track $\tau \subset F$ is called *recurrent* if for each branch b of τ, there is a multiple curve C_b carried by τ with supporting map $\phi: F \to F$ so that $b \subset \phi(C_b)$, or, equivalently, $\mu_{C_b}(b) > 0$. Some examples of recurrent and non-recurrent train tracks are given in Figure 1.3.1. (The verifications are routine and left as exercises.) The choice of terminology comes from the fact that the bi-infinite trainpath on τ associated with C_b passes through b infinitely often, so b "recurs" along this path. Conversely, suppose that $\rho: (-\infty, \infty) \to \tau$ is such a trainpath so that $p = \rho(r) = \rho(t) \in b$ and the tangents $\rho'(r)$ and $\rho'(t)$ have the same direction at p for some $r < t$. Thus, $\rho \mid_{[r,t]}$ is a closed trainpath on τ which traverses b, and the corresponding multiple curve C_b (see §1.2) satisfies the conditions above. Furthermore, it is not difficult to check that τ is recurrent if and only if it carries a multiple curve C with $\mu_C(b) > 0$ simultaneously for every branch b: indeed, if for each branch b, C_b is a multiple curve so that $\mu_{C_b}(b) > 0$, then the sum $\mu = \sum_b \mu_{C_b}$ satisfies the switch condition at each vertex (since the switch condition is linear), and μ determines an appropriate multiple curve C. Notice that if τ is an arbitrary train track, then there is a canonically defined *maximal recurrent subtrack* $\sigma \subset \tau$: a branch b of τ lies in σ if and only if there is a multiple curve C in F with $\mu_C(b) > 0$.

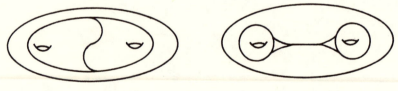

non-recurrent   recurrent

FIGURE 1.3.1

A *transverse measure* (or simply a *measure*) μ on τ is a function which assigns to each branch b of τ a nonnegative real number $\mu(b) \in \mathbb{R}_+ \cup \{0\}$

18

which satisfies for each switch v of τ the switch condition

$$\mu(e_1) + \ldots + \mu(e_r) = \mu(e_{r+1}) + \ldots + \mu(e_{r+t}),$$

where $e_1, \ldots e_r$ are the incoming ends of branches which are incident on v and $e_{r+1}, \ldots e_{r+t}$ are the outgoing ones (and the measure of an end is the measure of the corresponding branch, as before). The pair (τ, μ) is called a *measured train track*. If b is a branch of τ, then the number $\mu(b)$ is called the *weight on* (or *measure of*) b, and we write simply $\mu > 0$ if each branch has nonzero weight. A measure which is $[\mathbb{Z}_+ \cup \{0\}]$-valued is called an *integral (transverse) measure*.

An integral measure on a train track describes a multicurve in F (as in §1.2). More generally, when all the weights are rational, a measured train track describes a *weighted* multicurve, as we shall see. Still more generally, if the weights are not rationally related, then a measured train track describes a "measured geodesic lamination" (to be defined in §1.7).

Let n denote the number of branches of τ. Of course, the set of all $\mathbb{R}$-valued functions defined on the set of branches of τ is naturally identified with $\mathbb{R}^n$, and the collection of such functions which satisfy the switch conditions is a sub-vector space $H \subset \mathbb{R}^n$ since each switch condition is linear. The collection of all transverse measures on τ is then $H \cap [\mathbb{R}_+ \cup \{0\}]^n$, and the collection of all measures which are strictly positive is $V = H \cap \mathbb{R}_+^n$.

Now, if τ is recurrent, then it supports a measure $\mu > 0$ as we saw above. Conversely, if there is a measure $\mu > 0$ on τ, then V is not empty and is therefore homeomorphic to an open cell of the same dimension as H itself. Inside V, we may then find a point with rational coordinates; multiplying all weights to clear denominators thus gives an integral point in V, and this point describes a multiple curve $C \subset F$ with $\mu_C > 0$, so τ is recurrent. Thus we have proven

Proposition 1.3.1: *A train track τ is recurrent if and only if it supports a transverse measure $\mu > 0$.*

Suppose that σ is either another train track in F or perhaps a multiple curve in F which intersects τ transversely (if at all). We say that σ *hits τ efficiently* provided that no component of $F - \sigma - \tau$ is an imbedded bigon in F. A seemingly stronger characterization of hitting efficiently (which is quite useful in the sequel) is provided by

Proposition 1.3.2: *Suppose that $\tau \subset F$ is a train track and $\sigma \subset F$ is either another train track or perhaps a multiple curve which meets τ transversely (if at all), and let $\tilde{\sigma}$ and $\tilde{\tau}$, respectively, denote the full pre-images of σ and τ in the universal cover $\tilde{F}$ of F. Then σ hits τ efficiently if and only if there is no imbedded bigon in $\tilde{F}$ whose frontier is contained in $\tilde{\sigma} \cup \tilde{\tau}$.*

Proof: It is obvious that the stated condition is sufficient for σ and τ to hit efficiently. Conversely, suppose $B \subset \tilde{F}$ is such a bigon. By the first part of Corollary 1.1.2, it must be that one frontier edge, say E_σ, of B lies in $\tilde{\sigma}$ and the other, say E_τ, in $\tilde{\tau}$. Any trainpath ρ on $\tilde{\tau}$ is imbedded in $\tilde{F}$ by the second part of Corollary 1.1.2. Furthermore, since B has compact closure (by definition of an imbedded bigon) and $\tau \subset F$ is closed, the closure of each component of $\rho \cap B$ must meet the frontier of B in two distinct points. Any such trainpath which meets both E_τ and B must also meet E_σ, for otherwise there would be an imbedded monogon or bigon in $\tilde{F}$ whose frontier edges consist of finite trainpaths on $\tilde{\tau}$, in contradiction to the first part of Corollary 1.1.2. Thus, any trainpath on $\tilde{\tau}$ which meets B gives rise to an imbedded sub-bigon of B which has one frontier edge in $\tilde{\sigma}$ and one frontier edge in $\tilde{\tau}$. The analogous assertion also holds for $\tilde{\sigma}$, and it follows that there is an innermost such bigon, say D, with $D \cap (\tilde{\sigma} \cup \tilde{\tau}) = \emptyset$.

It remains only to show that D projects to an imbedded bigon in F. In the contrary case, there are distinct points $p, q \in D$ and a covering translation γ with $\gamma(p) = q$. If x lies in the frontier of D, then there is an arc α with endpoints p, x which is otherwise disjoint from $\tilde{\sigma} \cup \tilde{\tau}$. Since $q \in \gamma(\alpha)$ and $\gamma(\alpha) - \gamma(x)$ is also disjoint from $\tilde{\sigma} \cup \tilde{\tau}$, it follows that γ preserves the frontier of D and hence preserves the set of vertices of D. Thus, the composition $\gamma \circ \gamma$ fixes each vertex of D, which is absurd for a non-trivial covering translation.   q.e.d.

Dual to the notion of recurrence, τ is *transversely recurrent* if for each branch b of τ, there is a multiple curve C_b which hits τ efficiently so that C_b intersects b at least once. A train track which is both recurrent and transversely recurrent is called *birecurrent*, and these are the typical train tracks we consider in subsequent chapters. Each of the train tracks already considered in Figure 1.3.1 is transversely recurrent though only one of them is recurrent. We remark parenthetically that while recurrence is an "intrinsic" property of a track in the sense that it is independent of *how* the track is imbedded in the surface, transverse recurrence is an "extrinsic" property.

Two other conditions which are equivalent to transverse recurrence will be provided in Corollary 1.3.5 below (see also Theorem 1.4.3), and a certain duality between recurrence and transverse recurrence will be explored in §3.4. Two examples of train tracks which are recurrent but not transversely recurrent are provided in Figure 1.3.2; we leave the verification of this as an exhausting exercise using the definition above. (This verification is an easy consequence of Corollary 1.3.5; see the example following the proof of that result.)

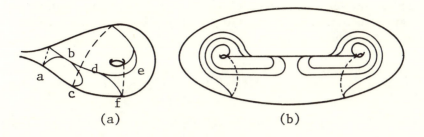

(a)   (b)

FIGURE 1.3.2

Lemma 1.3.3: (a) If τ' is a subtrack of the transversely recurrence track τ, then τ' is transversely recurrent as well.

(b) If τ is a transversely recurrent train track which carries a train track $\tau' < \tau$, then τ' is transversely recurrent as well.

Proof: For part (a), suppose that b' is a branch of τ'. Since $\tau' \subset \tau$, either the closure of b' contains the closure of a union of branches of τ or perhaps b' is a curve component C of τ' which is itself the closure of union of branches of τ. In the former case, b' contains a branch b of τ, and by transverse recurrence, there is some multiple curve C_b hitting τ efficiently with $C_b \cap b \neq \emptyset$. Thus, $C_b \cap b' \neq \emptyset$, and by Proposition 1.3.2, C_b also hits τ' efficiently. In the latter case, C similarly contains a branch of τ, and one easily produces the required curve. Since the branch b' of τ' was arbitrary, transverse recurrence of τ' is assured.

For part (b), let $\phi: F \to F$ denote the supporting map of the carrying $\tau' < \tau$, and choose a branch b' of τ'. Since $d\phi_p$ is nonzero on the tangent line to τ' at p for each $p \in \tau'$ (by Condition (3) in the definition of carrying), $\phi(b')$ contains some arc a which lies in the interior of a branch b of τ. Let C be a multiple curve in F meeting τ efficiently with $b \cap C \neq \emptyset$. By isotoping C near b, we can arrange that $a \cap C \neq \emptyset$, and by a further small isotopy if necessary, we can arrange that each point of $C \cap \tau$ is a regular value of ϕ. Thus, $C' = \phi^{-1}(C)$ is a curve in F with $b' \cap C' \neq \emptyset$, and in fact, C' is a multiple curve since it is homotopic to C.

If $B \subset F$ were a bigon component of $F - (\tau' \cup C')$, then $\phi(B) \subset F$ would be a bigon component of $F - (\tau \cup C)$, which is absurd, so C' meets τ' efficiently. Since b' was arbitrary, we conclude that τ' is transversely recurrent, as desired.   q.e.d.

Suppose that τ is a track and ν is an assignment of nonnegative real number $\nu(b) \in \mathbb{R}_+ \cup \{0\}$ to each branch b of τ. (We are *not* requiring that ν satisfy the switch conditions.) If $\rho: [r, t] \to \tau$ is a finite trainpath on τ, then ν determines a number

$$\nu(\rho) = \sum_{k=r}^{t-1} \nu(\rho(k, k+1)) \in \left[\mathbb{R}_+ \cup \{0\} \right]$$

in the natural way. Dual to the notion of a (transverse) measure on a track τ, a nonnegative function ν defined on the branches of τ is a *tangential measure* on τ provided that the following two conditions hold.

(1) Let D be an m-gon component of $F - \tau$ (so $m \geq 3$), whose frontier edges correspond to the finite trainpaths $\rho_1, \ldots, \rho_m$, respectively, on τ. We demand that

$$\nu(\rho_i) \leq \sum_{k \neq i} \nu(\rho_k), \text{ for each } i = 1, \ldots, m.$$

(2) Supppose that A is an m-gon-minus-a-disk component of $F - \tau$ (so $m \geq 1$), whose frontier edges correspond to the finite trainpaths $\rho_0, \rho_1, \ldots, \rho_m$, respectively on τ, where ρ_0 corresponds to the smooth frontier edge of A. We demand that

$$\nu(\rho_0) \leq \sum_{k=1}^{m} \nu(\rho_k).$$

Other complementary regions impose no further conditions. As with transverse measure, we write $\nu > 0$ if $\nu(b) > 0$ for each branch b of τ and say that ν is an *integral tangential measure* if it is $[\mathbb{Z}_+ \cup \{0\}]$-valued. If ρ is a finite trainpath on a track τ supporting a tangential measure ν, then the number $\nu(\rho)$ (described above) is called the *total (ν-)tangential measure* of ρ. Finally, an integral tangential measure is said to be *even* if the following condition holds for each component R of $F - \tau$: if $\rho_1, \ldots, \rho_m$, respectively, are the finite trainpaths on τ corresponding to the C^1 curves and arcs in the frontier of R, then the sum $\sum_{k=1}^{m} \nu(\rho_k)$ of total tangential measures is even.

Suppose that $C \subset F$ is a multiple curve hitting a train track $\tau \subset F$ efficiently, and suppose that R is a component of $F - \tau$. C meets the frontier of R an even number of times since the intersection $C \cap R$ consists of a collection of curves and arcs disjointly imbedded in R with the endpoints of the arcs in the frontier of R. If R is an m-gon component of $F - \tau$, then each such arc has its endpoints on distinct frontier edges of R since C hits τ efficiently. Similarly, if R is a component of $F - \tau$ which is an m-gon-minus-a-disk, then $C \cap R$ is again a collection of arcs disjointly imbedded in R, and each such arc either has one endpoint in the smooth component and one in the non-smooth component of the frontier of R, or perhaps has both endpoints in the non-smooth component. (Furthermore, if both endpoints of such an arc lie in a common frontier edge of R, then the arc decomposes R into an $(m+2)$-gon and a bigon-minus-a-disk.) It follows that if we define

$$\nu_C(b) = \text{ card } (C \cap b), \text{ for } b \text{ a branch of } \tau,$$

then ν_C is an even tangential measure on τ.

Remark: More generally, if C is a collection of closed one-manifolds immersed in F and $\tau \subset F$ is a train track, then we say C "hits τ efficiently" if there is no path homotopy (fixing the endpoints) between sub-arcs of C and smooth paths on the boundary of a component of $F - \tau$. In this case, there is an induced even tangential measure $\nu_C(b) = \text{ card } (C \cap b)$, for b a branch of τ, exactly as above.

A converse to the discussion above is given by

Lemma 1.3.4: *If ν is an even integral tangential measure on a train track $\tau \subset F$, then there is a multiple curve C hitting τ efficiently so that C intersects each branch b of τ in exactly $\nu(b)$ points.*

Proof: We may add simple closed curves (each containing exactly one bivalent switch) to τ to obtain a track τ' so that each component R of $F - \tau'$ is one of the following:

Case i): An m-gon, for $m \geq 3$.

Case ii): A once-punctured m-gon, for $m \geq 1$.

Case iii): An m-gon-minus-a-disk , for $m \geq 1$.

Case iv): A pair of pants, a twice-punctured nullgon, or a once-punctured annulus; in each of these circumstances, we refer to R simply as a *pseudo pair of pants*.

Some care is required to extend ν to an even tangential measure ν' on τ'. Consider adding the components $\{\delta_i\}_{i=1}^{I}$ of $\tau - \tau'$ one at a time, and suppose by induction that we have defined an even tangential measure ν_t on $\tau_t = \tau \cup \{\delta_i\}_{i=1}^t$, for $t \geq 0$. Let R_1 and R_2 be the components of $F - \tau_{t+1}$ on either side of δ_{t+1}. If $R_1 = R_2$, then we extend ν_t to ν_{t+1} with the definition $\nu_{t+1}(\delta_{t+1}) = 0$. If R_1 and R_2 are distinct, then let $\rho_0, \rho_1, \ldots \rho_K$, respectively, denote the finite trainpaths corresponding to the smooth curves and arcs in the frontier of R_1, where ρ_0 corresponds to δ_{t+1}, and define $\xi_1 = \sum_{k=1}^K \nu_t(\rho_k)$; similarly, let ξ_2 denote the sum of the total ν_t-tangential measures of trainpaths corresponding to the frontier of R_2 minus δ_{t+1}. Since $R_1 \cup \delta_{t+1} \cup R_2$ is a component of $F - \tau_t$, it follows that $\xi_1 + \xi_2$ is even by the inductive hypothesis. If both ξ_1 and ξ_2 are even, take $\nu_{t+1}(\delta_{t+1}) = 0$, and, if both are odd, take $\nu_{t+1}(\delta_{t+1}) = 1$. This completes the inductive step, and we take $\nu' = \nu_I$ as a tangential measure on $\tau' = \tau_I$ satisfying the required evenness conditions.

Before constructing the putative multiple curve C, we further modify the track τ' so that each complementary once-punctured m-gon and m-gon-minus-a-disk (cases ii) and iii) above) satisfy $m = 1$; given such a region R with $m > 1$, let ξ denote the sum of the total ν'-tangential measures of trainpaths corresponding to the non-smooth component of the frontier of R, and choose a switch v of τ' on this non-smooth component; add a new branch $b \subset R$ both of whose ends are incident on v as in Figure 1.3.3, and extend ν' to a measure on the track so constructed by taking the weight on b to be ξ. Performing this modification in each appropriate region of type ii) or iii) produces a track τ'' supporting an even tangential measure ν''.

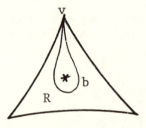

FIGURE 1.3.3

Since each branch of τ is a branch of $\tau'' \supset \tau$ and ν'' agrees with ν on these branches and since a multiple curve hitting τ'' efficiently must hit τ efficiently by Proposition 1.3.2, it suffices to prove the lemma for the tangential measure ν'' on τ''. To this end, let us pick $\nu''(b)$ distinct points on each branch b of τ''; we will construct an appropriate multiple curve C by connecting these points with arcs in each component R of $F - \tau''$ as follows. In case iv) so that R is a pseudo pair of pants, choose a collection of arcs in R as in Dehn's Theorem (see Figure 1.2.2) which meet the frontier of R in our chosen points. In case iii) so that R is a monogon-minus-a-disk, let v be the vertex in the frontier of R, let ξ_1 and ξ_2, respectively, denote the total tangential measures corresponding to the smooth and non-smooth components δ_1 and δ_2, respectively, of the frontier of R, and connect the $\frac{1}{2}(\xi_2 - \xi_1) \in \mathbb{Z}_+ \cup \{0\}$ points on δ_2 immediately to the left of v to the same number immediately to the right; the remaining ξ_1 points on δ_2 are then connected to the ξ_i points on δ_1. (See Figure 1.3.4a.) Case ii), in which R is a once-punctured m-gon, is handled similarly.

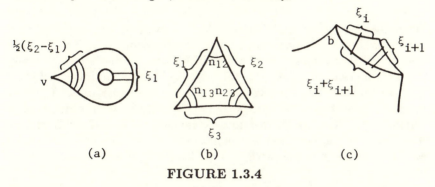

$$\text{FIGURE 1.3.4}$$

We are left to define the construction of arcs only in case i) so that R is an m-gon with $m \geq 3$, and we proceed by induction on m. First, suppose that $m = 3$ and let ξ_1, ξ_2, ξ_3, respectively be the total tangential measures corresponding to each of the frontier edges of R. The equations

$$\xi_i = \sum_{i \neq j = 1}^{3} n_{ij}, \text{ for } i = 1, 2, 3,$$

have a $unique$ nonnegative solution since the ξ_i are constrained by Condition (1) in the definition of tangential measure; namely

$$n_{ij} = \frac{1}{2}(\xi_i + \xi_j - \xi_k) \text{ for } \{i, j, k\} = \{1, 2, 3\},$$

and this solution is integral by the evenness condition. The basis step is completed by connecting the ith and j^{th} frontier edges of R with n_{ij} arcs as in Figure 1.3.4b, for distinct $i, j \in \{1, 2, 3\}$. For the inductive step, suppose that R is an m-gon with $m > 3$, and let $\xi_1, \ldots, \xi_m$, respectively, denote the total tangential measures of the frontier edges of R enumerated in the order in which they occur in a C^0 traversal of the frontier of R; let us also regard the indices as determined mod m, so that, for instance, $\xi_{m+1} = \xi_1$. Select $i \in \{1, \ldots, m\}$ so that $\xi_i + \xi_{i+1}$ is a minimum of $\{\xi_j + \xi_{j+1} : j = 1, \ldots, m\}$,

and add a branch b to τ'' separating the ith and $(i+1)^{\text{st}}$ frontier edges
from the rest as in Figure 1.3.4c, so that b decomposes R into a trigon and
an $(m-1)$-gon. Select $\xi_i + \xi_{i+1}$ points on the new branch b, and complete
in each of these regions using the inductive hypothesis to construct the
required collection of arcs in R.

By construction, the family of arcs produced above combine to give a
collection C of simple closed curves in F, and we can easily arrange that C
is C^1. Furthermore, C meets each branch b of τ'' in exactly $\nu''(b)$ points,
each component of C intersects τ'' transversely in a non-empty set, and
no component of $F - (C \cup \tau'')$ is a bigon. To complete the proof of the
lemma, we must show that C is a multiple curve (i.e., no component of C
is null-homotopic or puncture-parallel). If some component c of C is null-
homotopic, then c must bound a disk in F; since τ'' meets c transversely,
τ'' must intersect the interior of this disk, and we find an imbedded bigon
in F so that one frontier edge lies in τ'' and the other in c. As in the
proof of Proposition 1.3.2, there is an innermost such bigon, which must be
disjoint from $C \cup \tau''$ and is therefore a component of $F - (C \cup \tau'')$, which is
impossible. In the same way, no component of C can be puncture-parallel,
so C is in fact a multiple curve, and the lemma is proved.   q.e.d.

Remarks: 1) If each complementary region of τ in F is either a trigon
or a once-punctured monogon and ν is an even tangential measure on τ,
then there are essentially no choices in the construction above of a multiple
curve hitting τ efficiently and realizing $\nu(b) = \text{card}\, (C \cap b)$ for each branch
b of τ; the underlying multicurve is actually unique, as we shall see later.
However, if some component of $F - \tau$ is other than above, then this is not
true. Furthermore, in contrast to transverse measure, different tangential
measures on the same track can give rise to the same multicurve. These
points will be addressed in §3.4.

2) The last step in the argument above proves the general fact that if a
collection C of simple curves disjointly imbedded in F hits a train track in
F efficiently, then C is in fact a multiple curve.

An analogue of Proposition 1.2.2 for transversely recurrent tracks (see
also Theorem 1.4.3) is contained in the following

Corollary 1.3.5: *Suppose that $\tau \subset F$ is a train track. The following
conditions are equivalent:*

(i) τ is transversely recurrent.

(ii) τ supports an even (integral) tangential measure $\nu > 0$.

(iii) There is a multiple curve $C \subset F$ hitting τ efficiently so that $C \cap b \neq \emptyset$, for every branch b of τ.

Proof: By definition of transverse recurrence, we can find for each branch b of τ a multiple curve $C_b \subset F$ so that C_b hits τ efficiently and $b \cap C_b \neq \emptyset$. The multiple curve C_b determines an even tangential measure $\nu_b = \nu_{C_b}$ as before. The sum

$$\nu = \sum_{b \text{ a branch of } \tau} \nu_b$$

is a tangential measure on τ (by linearity of the inequalities in the definition of tangential measure) which is even, so (i) implies (ii). Furthermore, the multiple curve $C \subset F$ which arises by applying the previous lemma to ν meets each branch b of τ since $\nu(b) \neq 0$, so (ii) implies (iii). Since (iii) obviously implies (i), the proof is complete.   q.e.d.

Example: Armed with Corollary 1.3.5, we now verify that the train track $\tau \subset F = F_1^1$ in Figure 1.3.2a is not transversely recurrent. Adopt the notation in the figure for branches of τ, so that one component of $F - \tau$ is a once-punctured monogon and the other is a trigon, whose three frontier edges correspond to finite trainpaths on τ which traverse branches df, $edbabcf$, and ec, respectively. If τ were transversely recurrent, then by the previous result, it would support an even tangential measure $\nu > 0$. By Condition (1) in the definition of tangential measure, we must have $\nu(e) + \nu(d) + 2\nu(b) + \nu(a) + \nu(c) + \nu(f) \leq \nu(c) + \nu(e) + \nu(d) + \nu(f)$, so $\nu(a) = \nu(b) = 0$, contradicting that $\nu > 0$. This establishes that τ is not transversely recurrent. The verification that the track in Figure 1.3.2b is not transversely recurrent is similar and left as an exercise.

We say that a train track is *maximal* if it is not a proper subtrack of any other train track. Of course, we can add branches to any track which is not maximal until each complementary region is either a trigon or a once-punctured monogon (as in the proof of Proposition 1.1.1). We say that a birecurrent train track is *complete* if it is not a proper subtrack of any birecurrent track, and a more interesting problem involves extending a birecurrent train track to a complete train track.

Theorem 1.3.6: *(a) If $g > 1$ or $s > 1$, then any birecurrent train track on $F = F_g^s$ is a subtrack of a complete train track, each of whose complementary regions is either a trigon or a once-punctured monogon.*

(b) Any birecurrent train track on $F = F_1^1$ is a subtrack of a complete train track, whose unique complementary region is a once-punctured bigon.

Proof: To begin the construction of a complete birecurrent train track with a given birecurrent train track τ as subtrack, we can add curve components to τ as in the first step of the proof of Lemma 1.3.4 to produce a train track τ' so that each component of $F - \tau'$ is an m-gon, a once-punctured m-gon, an m-gon-minus-a-disk, or a pseudo pair of pants. Recurrence of τ' follows immediately from the recurrence of τ, and transverse recurrence of τ' is proved as follows. By Corollary 1.3.5, we may choose a strictly positive even tangential measure on τ which we may multiply by two to produce an even tangential measure ν so that $\nu(b) \geq 2$ for each branch b of τ. Let us now apply the inductive argument given at the beginning of the proof of Lemma 1.3.4 with a small difference: wherever we before extended ν by taking a tangential measure zero on an added curve, we extend this time by taking tangential measure 2. As before, this produces a strictly positive tangential measure on τ' which satisfies the hypotheses of Lemma 1.3.4, and application of this lemma finally proves that τ' is transversely recurrent, as was claimed.

We next proceed to add branches across complementary regions of τ', and some care is needed to guarantee that the result is birecurrent. We will add branches to τ' in a complementary region R in one of the following ways:

> **Move 1:** If R is an m-gon, $m > 3$, then add a branch in R connecting two vertices which are not consecutive along the frontier; see Figure 1.3.5(1).

> **Move 2:** If R is a once-punctured m-gon, $m > 1$, then add a branch in R encircling the puncture with both endpoints at a single vertex in the non-smooth component of the frontier of R; see Figure 1.3.5(2).

> **Move 3:** If R is an m-gon-minus-a-disk, $m \geq 1$, then add a branch in R which connects two switches of τ, one in each of the two components of its frontier; see Figure 1.3.5(3).

> **Move 4:** If R is a pseudo pair of pants, then add a branch in R connecting any two switches of τ lying in distinct components of the frontier of R; see Figure 1.3.5(4).

> **Move 5:** If R is a pseudo pair of pants, then add a branch e in

R running from a switch of τ in the frontier of R back to itself so that each component $R - e$ is either a once-punctured monogon or a monogon-minus-a-disk; see Figure 1.3.5(5).

Move 6: If R is a once-punctured m-gon, $m > 2$, and δ is a frontier edge of R with vertices v and w, then add a branch e in R connecting v to w so that the component of $R - e$ containing the puncture of R has e and δ as its frontier edges; see Figure 1.3.5(6).

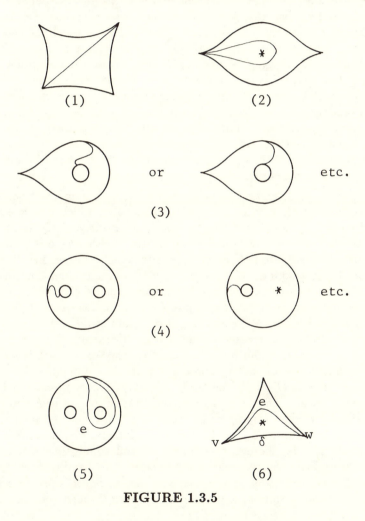

FIGURE 1.3.5

It is obvious that the application of any of these moves to a train track produces a train track with exactly one new branch, and we furthermore

claim that these moves preserve transverse recurrence. It certainly suffices
to consider the case that a train track σ' arises from a transversely recurrent
train track σ by adding only one branch, say $e \subset R$, where R is a component
of $F - \sigma$, in accordance with the moves; we must prove that σ' is transversely
recurrent as well. To this end, first note that if $b \neq e$ is a branch of σ' and
C is a multiple curve hitting σ efficiently with $C \cap b \neq \emptyset$, then an isotopy of
C fixing $F - R$ pointwise (preceded by a Dehn twist along a curve in $F - \sigma$,
if necessary, in case of Move 3 or 4) produces a multiple curve hitting σ'
efficiently and meeting b. The problem is to find a multiple curve meeting e
and hitting σ' efficiently. For Moves 4 and 5, a parallel copy of a component
of ∂R which contains (at least one of) the vertices on which the new branch
e is incident, meets e and hits σ' efficiently. For Move 3, the core circle of
the underlying topological annulus R similarly solves the problem. For
Move 2, we argue as in Lemma 1.3.4 as follows: since σ is transversely
recurrent, it supports a positive even integral tangential measure, say ν,
by Corollary 1.3.5; let ξ denote the sum of the total ν-tangential measures
of trainpaths corresponding to the non-smooth component of the frontier
of R, and extend ν to an even tangential measure ν' on σ' by defining the
weight of e to be ξ; finally, application of Lemma 1.3.4 to ν' produces the
required multiple curve.

For Move 1, choose branches $e_i, i = 1, 2$, of σ in the frontier of R
on opposite sides of e in R. Let C_i be a curve which hits σ efficiently
and crosses e_i, for $i = 1, 2$. As before, an isotopy of C_i fixing $F - R$
pointwise arranges that C_i hits σ' efficiently. If either C_1 or C_2 now meets
e, our argument is complete, so suppose that neither does. Let R_1 and R_2,
respectively, denote the components of $R - e$ whose frontier contains e_1 and
e_2. There are three possibilities to consider: either C_1 meets R_2, C_2 meets
R_1, or neither curve meets the opposite region. When C_1 meets R_2, choose
an imbedded arc $\alpha \subset R$ connecting two points of C_1 and intersecting C_1
nowhere else, which crosses e exactly once. Construct C by surgering C_1
along α; C meets e and hits σ' efficiently. The second case is analogous.
In the final situation, again choose $\alpha \subset R$ meeting e in one point, but this
time α connects C_1 to C_2, meets $C_1 \cup C_2$ only in its endpoints, and crosses
e exactly once. Form a closed one-manifold C immersed in F which meets
e and hits σ' efficiently by surgering $C_1 \cup C_2$ along α. As in the remark
preceeding Lemma 1.3.4, C determines an integral tangential measure on
σ' which satisfies the evenness condition, and application of that lemma
finally produces the required multiple curve. To handle Move 6, choose
branches $e_i, i = 1, 2$ of σ, where $e_1 \subset \delta$ and e_2 is contained in a frontier
edge of R other than δ, and proceed *verbatim* through the argument for
Move 1.

In order to see that judicious application of the moves preserves recurrence, we introduce the notion of oriented tracks. An *orientation* on a

train track σ is a specification of orientation for each branch of σ which is consistent in the following sense: suppose that v is a switch and b_1, b_2 are branches of σ, and $f:(-1,1) \to \sigma$ is a C^1 imbedding with $f((-1,0)) \subset b_1$, $f(0) = v$, and $f((0,1)) \subset b_2$; if $f \mid_{(-1,0)}$ agrees with the prescribed orientation on b_1, then $f \mid_{(0,1)}$ agrees with the prescribed orientation on b_2. In other (less precise but more vivid) words, at each switch of σ, incoming points towards outgoing or vice-versa. A train track which admits an orientation is said to be *orientable*; otherwise, the track is *non-orientable*. Notice that if σ has a complementary region R so that some component of the frontier of R consists of m frontier edges (e.g., if R is an m-gon), where m is odd, then σ is necessarily non-orientable; the converse, however, is false; see Figure 1.3.6 for a counter-example.

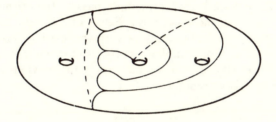

FIGURE 1.3.6

Proposition 1.3.7: *Suppose that σ is a connected recurrent train track and e, e' are (not necessarily distinct) branches of σ. Then we have*

> **Basic Fact 1:** *If σ is oriented, then there is a finite trainpath $\rho: [0, n] \to \sigma, n > 1$, with $\rho((0, 1)) = e$ and $\rho((n-1, n)) = e'$ so that the orientation of $\rho \mid_{(k,k+1)}$ agrees with the prescribed orientation on the branch $\rho((k, k+1))$ for each $k = 0, \ldots n - 1$.*

> **Basic Fact 2:** *If σ is non-orientable and the branches e and e' are each given an arbitrary orientation, then there is a finite trainpath $\rho: [0, n] \to \sigma, n > 1$ so that $\rho((0, 1)) = e$, $\rho((n-1, n)) = e'$ and the orientations of $\rho \mid_{(0,1)}$ and $\rho \mid_{(n-1,n)}$, respectively, agree with the prescribed orientations on e and e'.*

Proof: To prove the first basic fact, fix the branch e and let E denote the closure of the union of branches e' admitting such a train path starting from e. Suppose that $b_1 \subset E$ contains a switch v of σ in its closure and the specified orientation on b_1 corresponds to the incoming (outgoing, respectively) direction at v; if b_2 is another branch of σ so that the closure of b_2 contains v and the specified orientation on b_2 corresponds to the outgoing (incoming, respectively) direction at v, then we must have $b_2 \subset E$ as well. Suppose, to reach a contradiction, that there is some branch $b \subset \tau - E$ whose closure contains some point of E. By recurrence, there is some connected multiple curve C carried by σ with $\mu_C(b) \neq 0$, and C inherits an orientation from the orientation of σ. Since the closure of b contains a point of E, $\mu_C(b') \neq 0$ for some branch $b' \subset E$ by the observation above, and one easily produces an appropriate trainpath on σ which begins with e and ends with b. Thus, in fact $b \subset E$, so there can be no branch of $\sigma - E$ containing any point of E in its closure. By connectivity of σ, we must therefore have $E = \sigma$, as desired.

To prove the second basic fact, simply pass to the orientable double cover $\hat{\sigma}$ of σ (defined in the usual way). Select lifts $\hat{e}$ and $\hat{e}'$, respectively, of e and e' to $\hat{\sigma}$ so that $\hat{e}$ projects to e with the specified orientation, as does $\hat{e}'$ to e'. Recurrence of σ implies recurrence of $\hat{\sigma}$, and non-orientability of σ implies connectivity of $\hat{\sigma}$, so we may apply the previous argument to $\hat{\sigma}, \hat{e}, \hat{e}'$ in order to derive second basic fact.   q.e.d.

Armed with the notion of orientability and our knowledge that the Moves 1-6 preserve transverse recurrence, let us return to the extension of the birecurrent track τ' to a complete birecurrent track. Recall that each component of $F - \tau'$ is an m-gon, a once-punctured m-gon, an m-gon-minus-a-disk, or a pseudo pair of pants. Let us build an abstract finite connected graph G as follows: G has one vertex for each component of τ' and one vertex for each component of $F - \tau'$ which is an m-gon-minus-a-disk or a pseudo pair of pants; vertices of the former type are "solid", and vertices of the latter type are "hollow". Suppose that o is a hollow vertex corresponding to the complementary region R; to each end (in the sense of Carathéodory) of $R \cup \Delta$ (recall that Δ denotes the collection of distinguished points of $\hat{F}$, where $F = \hat{F} - \Delta$), there corresponds a (not necessarily distinct) component of τ' with corresponding solid vertex •, and we add to G one edge connecting o to • for each such end of R. Adding all such edges for each hollow vertex, we finally arrive at the graph G; see Figure 1.3.7 for an example, where we have moreover labeled each solid vertex with a sign $\pm$ to indicate whether it is orientable (and + designates orientability and − non-orientability). Each edge of G has one solid and one hollow endpoint by construction.

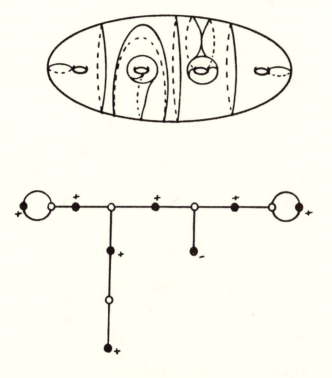

FIGURE 1.3.7

We collect in this paragraph several observations about univalent vertices of such a graph G. Notice first that each variety R of pseudo pair of pants, *except* the twice-punctured nullgon, has at least two ends in $R \cup \Delta$, and furthermore, each m-gon-minus-a-disk has two ends: it follows that a hollow univalent vertex of G must represent a twice-punctured nullgon. Suppose next that $\hat{\tau} \subset \tau'$ corresponds to a solid univalent vertex of G and that R is the m-gon-minus-a-disk or pseudo pair of pants component of $F - \tau'$ whose closure meets $\hat{\tau}$. We claim that either $\hat{\tau}$ contains the frontier of a (perhaps once-punctured) m-gon component of $F - \tau'$, or else R is an m-gon-minus-a-disk and $\hat{\tau}$ contains the non-smooth component of the frontier of R. (We do not claim that these possibilities are mutually exclusive.) Indeed, by univalence, $\hat{\tau}$ cannot be a curve component of τ', so $\hat{\tau}$ contains an r-valent switch, say v, where $r \geq 3$. If $\hat{\tau}$ does not contain the frontier of a (perhaps once-punctured) m-gon, then each region of $F - \tau$ near v lies in R by univalence. Since $\hat{\tau}$ contains an r-valent switch with $r \geq 3$, there must be a non-smooth point in the frontier of R. Thus, R is not a pseudo pair of pants, and our claim is proved.

Choose an orientation on each orientable component of τ' once and for all. Consider first an oriented imbedded edge-loop ℓ in G passing through solid vertices corresponding to the components $\tau_0, \tau_1, \ldots, \tau_r = \tau_0$ in this order. The loop ℓ must pass through the same number of hollow vertices, and we add a branch to τ' across each corresponding complementary region using Moves 3 and 4 to produce a train track τ'_ℓ; we must be careful to insure that the following conditions hold on the added branch b_{i-1} connecting τ_{i-1} to τ_i, for each $i = 1, \ldots, r$:

-If τ_i is oriented, then there is C^1 imbedding $f : (-1, 1) \to \tau_i \cup b_{i-1}$ with $f((-1, 0)) \subset b_{i-1}$ and $f((0, 1)) \subset \tau_i$ so that $f \mid_{(0,1)}$ agrees with the orientation on τ_i.

- In the same way, if τ_{i-1} is oriented, then there is likewise a C^1 imbedding $g : (-1, 1) \to \tau_{i-1} \cup b_{i-1}$ with $g((-1, 0)) \subset \tau_{i-1}$ and $g((0, 1)) \subset b_{i-1}$ so that $g \mid_{(-1,0)}$ agrees with the orientation on τ_{i-1}.

Since we have some latitude in the application of Moves 3 and 4 (as in Figures 1.3.5(3) and (4)), it is easy to satisfy these conditions.

The train track τ'_ℓ is transversely recurrent, as we showed above. We claim that τ'_ℓ is in fact also recurrent. To see this, first note that each point of $\tau' \subset \tau'_\ell$ is contained in the maximal recurrent subtrack of τ'_ℓ since τ' is itself recurrent. To finish the proof of our claim, we produce a closed trainpath ρ_ℓ on τ'_ℓ, which passes through each branch of $\tau'_\ell - \tau'$; this closed trainpath gives rise to a multiple curve carried by τ'_ℓ (as in §1.2), so each added branch lies in the maximal recurrent subtrack of τ'_ℓ as well, and τ'_ℓ is recurrent.

Let us suppose that the subscripts are taken mod r, so, for instance $b_r = b_0$. To begin the construction of ρ_ℓ, orient (the closure of) each branch b_{i-1} starting on τ_{i-1} and terminating on τ_i, for $1, \ldots, r$. (If $r = 1$ so that $\tau_1 = \tau_0$, orient b_0 arbitrarily.). By definition of Moves 3 and 4, the terminal switch of b_0 is a switch of τ_1, as is the initial switch of b_1. Furthermore, τ'_ℓ supports a trainpath of length two which first traverses b_0 (in its specified orientation) and then traverses some branch, say e_0, of τ_1; this trainpath induces an orientation on e_0 in the natural way. Similarly, there is a branch e_1 of τ_1 with an induced orientation and a trainpath of length two on τ'_ℓ which first traverses e_1 and then traverses b_1 (in its specified orientation). If τ_1 is non-orientable, then by Basic Fact 2, it supports a trainpath, say ρ_1, which begins with e_0 and ends with e_1 in their induced orientations, and it follows that τ'_ℓ supports a finite trainpath which begins by traversing b_0, then traverses ρ_1, and finally traverses b_1. If τ_1 is oriented, then because of our care in adding branches, the orientations induced above on e_0 and e_1 agree with the specified orientations. By Basic Fact 1, τ_1 supports a finite trainpath which begins with e_0 and ends with e_1 with these orientations, and we produce the required finite trainpath on τ'_ℓ as before.

Now, to construct the closed trainpath ρ_ℓ, we begin with the oriented branch b_0. From the remarks above, we can continue ρ_ℓ along τ_1 and then traverse the oriented branch b_1. By the same argument, ρ_ℓ can be continued along τ_2 and then traverse b_2. Continuing in this way, we ultimately produce a closed trainpath ρ_ℓ on τ'_ℓ, completing the proof that τ'_ℓ is birecurrent.

Proceeding in this way, we may reduce to the case that the graph G corresponding to the train track τ' is a tree. Our next goal is to further reduce to the case that G consists of a single solid vertex, so that τ' is connected and each complementary region is a (perhaps once-punctured) m-gon. To this end, define the "type" of a vertex of G to be the specification of solid or hollow together with a determination of sign $\pm$ for a solid vertex, as above. If G contains more than one vertex, then there is an oriented imbedded edge-path π on G which begins and ends with distinct univalent vertices. There are various possibilities to consider depending on the pair of types of the endpoints of π; the relevant possibilities are enumerated as follows: $(\bullet^+, \bullet^+), (\bullet^+, \bullet^-), (\bullet^-, \bullet^-), (\circ, \circ), (\circ, \bullet^+)$, and $(\circ, \bullet^-)$. In each case, we will add branches to τ' using Moves 3 and 4 across regions corresponding to hollow vertices contained in the *interior* of π, being careful to add branches compatibly with the orientations of any oriented tracks encountered (as before). Let τ'_π denote the train track so constructed. As above, τ'_π is transversely recurrent and each point of $\tau' \subset \tau'_\pi$ lies in the maximal recurrent subtrack of τ'_π.

We first consider the case in which the endpoints of π are of type $(\bullet^-, \bullet^-)$, and let τ_0 and τ_1, respectively, denote the (non-orientable) components of τ' corresponding to the initial and final vertices of π. Just as before, there are oriented branches e_0 of τ_0 and e_1 of τ_1 and a trainpath ρ_π which traverses each branch of $\tau'_\pi - \tau'$ and extends to a trainpath that begins with e_0 and ends with e_1. Since τ_0 is non-orientable, Basic Fact 2 is applicable and implies that τ_0 supports a finite trainpath ρ_0 which ends with e_0 and begins with its reverse; similarly, τ_1 supports a trainpath ρ_1 which begins with e_1 and ends with its reverse. Define a trainpath on τ'_π by traversing ρ_0, then ρ_π, then ρ_1, and finally traversing the reverse of ρ_π. This trainpath is closed, so each branch of $\tau'_\pi - \tau'$ lies in the maximal recurrent subtrack of τ'_π. We conclude as before that τ'_π is in fact recurrent.

Next, suppose that the endpoints of π have type $(\bullet^-, \circ)$; as we proved in our remarks on univalent vertices above, the hollow endpoint of π corresponds to a twice-punctured annulus. In this region, we perform one application of Move 5 and argue as in the previous paragraph, where a trainpath of length one corresponding to this added branch plays the role of one of the trainpaths ρ_1 above. The case $(\circ, \circ)$ is handled similarly.

Suppose next that the endpoints of π have type $(\bullet^+, \bullet^-)$, and let τ_+ and τ_-, respectively, denote the oriented and non-orientable components of τ' corresponding to the endpoints of π. The usual argument produces oriented branches e_+ of τ_+ (where e_+ is oriented compatibly with τ_+) and e_- of τ_- and a finite trainpath ρ_π on τ'_π traversing each branch of $\tau'_\pi - \tau'$ so that ρ_π extends to a trainpath on τ'_π that begins with e_+ and ends with e_-. By Basic Fact 2, τ_- supports a trainpath ρ_- that begins with e_- and ends with its reverse.

Let us first assume that τ_+ contains the frontier of some complementary m-gon R. Since τ_+ is oriented, m is even, and we add an edge $e \subset R$ to τ_+ in accordance with Move 1 to produce a track τ''_π in such a way that e connects two terminal points of branches in the frontier of R (which are oriented compatibly with τ_+); see Figure 1.3.8a. Orient e arbitrarily and let w denote the initial switch of e. It follows from two applications of Basic Fact 1 that τ_+ supports a trainpath ρ_1 that begins by traversing e and ends by traversing e_+ and a trainpath ρ_2 that begins with the reverse of e_+ and ends at w. Finally, define the required closed trainpath on τ''_π as follows: begin by traversing ρ_1, then ρ_π, then ρ_-, then the reverse of ρ_π, and finally ρ_2. We add parenthetically that we could equally as well have chosen the branch e above to connect two *initial* points of branches in the frontier of R; to see that this is the case, simply reverse the orientation on τ_+.

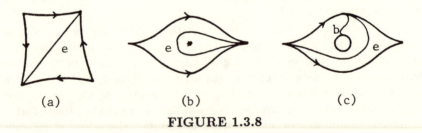

(a)   (b)   (c)

FIGURE 1.3.8

In case τ_+ contains the frontier of some complementary once-punctured m-gon, there is a similar argument using Move 2; see Figure 1.3.8b. In light of the results proved above about solid univalent vertices of G, to complete the case $(\bullet^+, \bullet^-)$, it remains only to consider the possibility that τ_+ bounds an m-gon-minus-a-disk R and contains the non-smooth component of the frontier of R. Again, since τ_+ is oriented, we must have $m \geq 2$ even, and the first branch b traversed by the trainpath ρ_π above (arising from an application of Move 3) decomposes R into an (m+2)-gon R'. In this case, we can add a branch e in R' using Move 1 with both its endpoints lying in τ_+, and argue analogously; see Figure 1.3.8c.

We are left to consider only the case $(\bullet^+, \bullet^+)$, and we let τ_1 and τ_2, respectively, denote the components of τ' corresponding to the initial and terminal vertices of the edge-path π. The usual argument produces a finite trainpath ρ_π which begins at a switch of τ_1 and ends at a switch of τ_2. Furthermore, at the expense of adding branches using Moves 1-3, arguments given in the previous case produce trainpaths ρ_i on τ_i, $i = 1, 2$, so that the first branch traversed by ρ_i is the reverse of the last branch traversed by ρ_i, and ρ_1, ρ_2, and ρ_π combine, as before, to produce the required closed trainpath traversing all the added branches.

Finally, we must consider the case where G is a single (necessarily solid) vertex, so that τ' is connected and each component of $F - \tau'$ is an m-gon, which is perhaps once-punctured. If τ' is non-orientable, then arrange that each puncture is contained in a once-punctured monogon by repeatedly using Move 2, and finally use Move 1 to arrange that the other complementary regions are trigons; as above, the resulting track is transversely recurrent, and Basic Fact 2 is easily applied after adding each branch to show that the result is recurrent as well. If τ' is orientable, then we may repeatedly apply Move 6 (which preserves orientability) and use Basic Fact 1 to show that recurrence is preserved as well until each puncture is contained in a once-punctured bigon. We may then use Move 1 *taking care to preserve orientability* and use Basic Fact 1 to preserve recurrence until each non-punctured complementary region is a fourgon. An easy computation shows that if κ and λ, respectively, are the number of complementary fourgons and once-punctured bigons, then $\kappa + \lambda = -\chi(F)$. In case $F = F_1^1$, we have constructed a track as in part (b) of Theorem 1.3.6. When F is not homeomorphic to this surface, there must be at least two distinct such complementary regions, say R_1 and R_2. Form a train track by adding one branch in each of R_1 and R_2 using Move 1 or 2 subject to the constraint that the branch added in R_1 (R_2, respectively) should connect two initial (terminal) points of oriented edges. Of course, the resulting track τ'' is transversely recurrent, and using Basic Fact 1, one easily proves as before that it is recurrent as well. Since τ'' has a complementary trigon, it is non-orientable, and the extension is now pursued as above.

It remains only to show that if $\sigma \subset F_1^1$ is a train track so that $F_1^1 - \sigma$ consists of exactly one once-punctured bigon R, then there is no birecurrent proper supertrack of σ in F_1^1. Indeed, we will show that there is not even a *recurrent* proper supertrack. To this end, suppose that $\sigma' \supset \sigma$ where the containment is proper. If b is a branch of $\sigma' - \sigma$, then $b \subset R$, and since σ' is a train track, there are various possibilities indicated in Figure 1.3.9. Thus,

$\sigma' - \sigma$ consists of the single branch b, and σ is easily seen to be orientable (as in Figure 1.3.9), while $\sigma' = \sigma \cup b$ is not. It follows that there can be no bi-infinite trainpath on σ' which traverses b more than once, so σ' is not recurrent, as asserted.

   This concludes the proof of Theorem 1.3.6.   q.e.d.

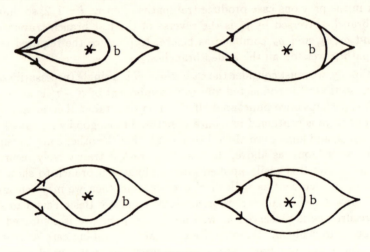

FIGURE 1.3.9

Remark: The construction of a complete supertrack of a given track in Theorem 1.3.6 suffers from the deficiency that the valence of switches of the supertrack might become quite large. This deficiency will be remedied in an important special case in the next section.

§1.4 GENERICITY AND TRANSVERSE RECURRENCE

We say that a train track τ is *generic* provided that each switch of τ is (either bivalent or) trivalent. Suppose that v is a trivalent switch of the generic track $\tau \subset F$ with incident half-branches $\hat{b}_1, \hat{b}_2, \hat{b}_3$, where every C^1 arc in τ through v meets $\hat{b}_1$; see Figure 1.4.1a. If $C \subset F$ is a multiple curve hitting τ efficiently so that $C \cap \hat{b}_1 \neq \emptyset$, then one can modify C by an isotopy supported in a neighborhood of $\hat{b}_1$ to produce a multiple curve in F which meets both $\hat{b}_2$ and $\hat{b}_3$ and still hits τ efficiently; see Figure 1.4.1b. This procedure is called *sneaking up* the multiple curve C, and we shall find it most useful. Indeed, the results of this section depend crucially on the availability of this technique. There is only the following weak analogue in the setting of non-generic train tracks: suppose that v is a switch of a train track $\tau \subset F$ and $C \subset F$ is a multiple curve which hits τ efficiently and meets *all* the half-branches corresponding to the incoming (or *all* the outgoing, respectively) ends incident on v; C may then be isotoped across v to produce a multiple curve which hits τ efficiently and meets all the half-branches corresponding to the outgoing (incoming) ends incident on v. This weak analogue is not sufficient for the arguments in this section.

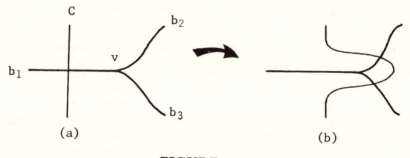

FIGURE 1.4.1

There is a modification (which we employed in the proof of Corollary 1.1.3) that one can perform on a non-generic track $\tau \subset F$ as follows.

39

Suppose that v is an r-valent switch of τ, where $r > 3$, let U be a closed regular neighborhood of v in F, and let $\hat{b}_1$ and $\hat{b}_2$ denote half-branches of τ which correspond to ends which are either both incoming or both out-going at v. Moreover, suppose that $\hat{b}_1$ and $\hat{b}_2$ are consecutive (i.e., when traversing the boundary ∂U of U, one consecutively encounters these half-branches); see Figure 1.4.2a. We may alter τ in U by identifying the arcs $\hat{b}_1 \cap U$ and $\hat{b}_2 \cap U$ to produce a train track τ', where the switch v gives rise to an $(r-1)$-valent switch $v' \in U$, the points $\hat{b}_1 \cap \partial U$ and $\hat{b}_2 \cap \partial U$ give rise to a trivalent switch $w \in \partial U$, and the identified arcs give rise to a branch $b' \subset U$ of τ'; see Figure 1.4.2b. The train track τ' is said to arise from τ by *combing*, and it is obvious that by repeated combing one can derive a generic track from a non-generic one.

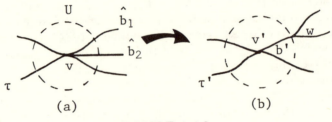

FIGURE 1.4.2

Proposition 1.4.1: *Suppose that τ is a non-generic train track and τ' arises from τ by combing.*

(a) τ is recurrent if and only if τ' is recurrent.

(b) If τ' is transversely recurrent, then τ is transversely recurrent as well.

Proof: It suffices to consider the case that τ' arises from τ by a single comb, and we adopt the notation above, where b_1 and b_2, respectively, denote the branches of τ' other than b' which are incident on w (and perhaps $b_1 = b_2$). There is a natural identification between the set of branches of τ and the set of branches of τ' other than b'. If τ' is recurrent, then it supports a measure $\mu' > 0$ by Proposition 1.3.1, and the definition $\mu(b) = \mu'(b)$, for b a branch of τ, produces a measure $\mu > 0$ on τ, so τ is recurrent by another application of Proposition 1.3.1. Furthermore, if τ' is transversely recurrent, then it supports an even tangential measure $\nu' > 0$ by Corollary 1.3.5, and the definition

$$\nu(b) = \begin{cases} \nu'(b) + \nu'(b'), & \text{if } b \in \{b_1, b_2\}; \\ \nu'(b), & \text{otherwise}, \end{cases}$$

for b a branch of τ, produces an even tangential measure $\nu > 0$ on τ, so τ is transversely recurrent again by Corollary 1.3.5.

To prove the converse in part (a), suppose that τ is recurrent, so it supports a measure $\mu > 0$ by Proposition 1.3.1. Define a measure $\mu' > 0$ on τ' by taking

$$\mu'(b) = \begin{cases} \mu(b_1) + \mu(b_2), & \text{if } b = b'; \\ \mu(b), & \text{otherwise}, \end{cases}$$

for b a branch of τ'. A final application of Proposition 1.3.1 shows that τ' is recurrent, and the proposition is proved.   q.e.d.

Remark: In fact, it is clear that if τ combs to τ', then $\tau' > \tau$, so part (b) above follows from Lemma 1.3.3b.

Unfortunately, the converse of part (b) above is not true as we show in the following example.

Example: Let $\tau \subset F_2^0$ be the non-generic track indicated in Figure 1.4.3, and adopt the notation there for the branches of τ. One can easily check that an assignment $\nu : \{a, \ldots, g\} \to 2\mathbb{Z}_+$ defines a tangential measure if and only if $\nu(d) = \nu(c) + \nu(b)$ and $\nu(g) = \nu(e) + \nu(f)$, so certain of the triangle inequalities are equalities. As ν is even since it has image in $2\mathbb{Z}$, τ is transversely recurrent by Corollary 1.3.5. On the other hand, it is an exercise to show that τ combs to the train track indicated in Figure 1.3.2b, which is *not* transversely recurrent. Thus, the converse to part (b) above is false, as was asserted. We add parenthetically that since τ combs to a recurrent track, it is recurrent by Proposition 1.4.1, part (a).

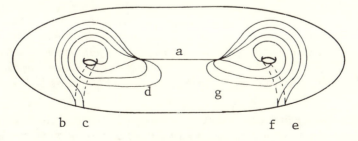

FIGURE 1.4.3

As promised before, we now address the deficiencies in the proof of Theorem 1.3.6.

Corollary 1.4.2: *(a) If $g > 1$ or $s > 1$ then any birecurrent generic train track on $F = F_g^s$ is a subtrack of a complete generic train track, each of whose complementary regions is either a trigon or a once-punctured monogon.*

(b) Any birecurrent generic train track on $F = F_1^1$ is a subtrack of a complete generic train track, whose unique complementary region is a once-punctured bigon.

Proof: The proof follows closely that of Theorem 1.3.6, but we must modify Moves 1-6 to preserve genericity. In each case, simply perform the move as before, and then comb the result to a generic track. It is easy to see that these modified moves preserve transverse recurrence using the technique of sneaking up; indeed, this step of the argument is simpler than before. The modified moves are then applied just as before to produce the desired complete generic track.   q.e.d.

Most of this section is devoted to a proof of the following theorem, which includes an analogue for transverse recurrence of Proposition 1.3.1 in the setting of generic tracks.

Theorem 1.4.3: *The following conditions on a generic train track $\tau \subset F$ are equivalent:*

(i) τ is transversely recurrent.

(ii) τ supports a tangential measure $\nu > 0$ satisfying strict inequalities in Conditions (1) and (2) in the definition of tangential measure.

(iii) For every $\varepsilon > 0$ and $L > 0$, there is a hyperbolic structure on F (i.e., a complete Riemannian metric of finite-area with constant Gauss curvature -1) so that τ has geodesic curvature less than ε at each point, and each branch of τ has length at least L.

Proof: Let us first prove that (i) implies (ii). Since τ is transversely recurrent, there is some multiple curve $C \subset F$ hitting τ efficiently and meeting each branch of τ by Corollary 1.3.5, and there is a corresponding tangential measure $\nu = \nu_C > 0$ on τ. Suppose that R is some m-gon component of $F - \tau$ so that equality holds for ν on R in Condition (1) in the definition of tangential measure. It follows that $C \cap R$ consists of a collection of arcs imbedded in R which all have exactly one endpoint on some common frontier edge, say δ, of R, and we choose a vertex v of R which is not an endpoint of δ. Since v is a trivalent switch by genericity of τ, there is a unique branch, say b, of τ which contains points of any C^1 arc through v, and by construction, $b \cap C \neq \emptyset$. Consider the component c of C which meets b nearest v, add a parallel copy c' of c near c, and sneak up c' across v to produce a new multiple curve $C' = C \cup \{c'\} \subset F$. See Figure 1.4.4a. The corresponding tangential measure $\nu_{C'}$ on τ satisfies the strict inequality on R in Condition (1) of the definition of tangential measure.

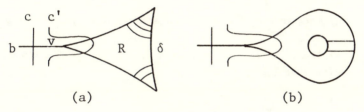

FIGURE 1.4.4

The case in which a tangential measure satisfies equality on an m-gon-minus-a-disk in Condition (2) of the definition of tangential measure is treated similarly; see Figure 1.4.4b. Proceeding in this way for each component of $F - \tau$ for which equalities hold in Conditions (1) and (2), we ultimately produce a multiple curve which meets τ efficiently and whose corresponding tangential measure satisfies the required conditions.

Next, we prove that Condition (ii) above implies Condition (i), and we proceed in analogy to the proof of Proposition 1.3.1. Suppose that τ has n branches and identify $\mathbb{R}^n$ with the set of all $\mathbb{R}$-valued functions defined on the set of branches of τ. The collection of such functions which satisfy the weak inequalities given in the definition of tangential measure form an intersection $H \subset \mathbb{R}^n$ of closed half-spaces. The set of tangential measures on τ is thus identified with $H \cap [\mathbb{R}_+ \cup \{0\}]^n$, and the set of strictly positive tangential measures is likewise identified with $V = H \cap \mathbb{R}_+^n$.

Now, since τ supports a strictly positive tangential measure which satisfies the strict inequalities in Conditions (1) and (2) in the definition of tangential measure, it follows that V has interior in $\mathbb{R}^n$, so we may choose a point in V with rational coordinates. Clearing denominators gives an integral point and finally multiplying by two (if necessary) produces an even tangential measure on τ. Corollary 1.3.5 then guarantees transverse recurrence of τ.

We next show that Condition (ii) implies Condition (iii). The proof is rather involved, requiring certain estimates, and was kindly contributed by Nat Kuhn. By Corollary 1.4.2 above, τ is a subtrack of some complete generic track, and if the supertrack can be imbedded in F with small geodesic curvature for some hyperbolic structure, then certainly so can τ. Thus, we may assume that τ is complete generic. The main point of the proof is summarized in

Theorem 1.4.4: *Assume that τ is transversely recurrent. Given positive constants L_0 and ε_0, there is a hyperbolic structure on F and a CW complex $T \subset F$ which is isomorphic to the underlying CW complex of τ so that T is piecewise geodesic and satisfies the following estimates:*

(a) *Each edge of T is a geodesic segment of length exceeding L_0.*

(b) *At each vertex of T, the angle of deviation from a straight line is less than ε_0; see Figure 1.4.5.*

FIGURE 1.4.5

Controlling L_0 and ε_0 is a discrete analogue of controlling the geodesic curvature, since geodesic curvature is a measure of angle of deviation per unit length. The following exercise completes the proof that Condition (ii) implies Condition (iii) given Theorem 1.4.4.

Exercise: Given $\varepsilon > 0$ and $L > 0$, there exist L_0 sufficiently large and ε_0 sufficiently small that a piecewise geodesic CW complex T satisfying (a) and (b) can be smoothed to have geodesic curvature less that ε and branches of length at least L. [Hint: Because of (a), there are disjointly imbedded neighborhoods U_v around each vertex v of T, so that each component of $U_v \cap (T - v)$ has length $\frac{1}{2}L_0$. Such a thing can be smoothed in a variety of ways.]

Proof of Theorem 1.4.4: For simplicity, let us first assume that F is compact, so that each complementary region of τ is a trigon; the non-compact case will be covered later. Let ν be a positive even tangential measure on τ as in Condition (ii). The strategy is to create a singular hyperbolic structure in which each branch b of τ has length $L\nu(b)$, where $L \gg 0$. Given a complementary trigon with frontier edges A, B, and C, the strict triangle inequalities guarantee the existence of a hyperbolic triangle with side lengths $L\nu(A), L\nu(B)$, and $L\nu(C)$. Break up each side of the triangle into segments of length $L\nu(b)$ corresponding to the branches b of τ in the frontier of the trigon; then glue together segments to get a geometric structure on F. Each vertex of a triangle comes from a "sector" of $F - \tau$ about a switch. Number the switches $1, 2, \ldots, N$, and let θ_i denote the interior angle at the vertex corresponding to the ith sector; see Figure 1.4.6a.

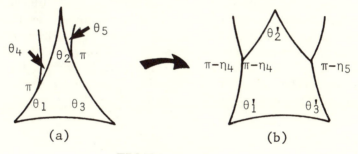

FIGURE 1.4.6

In the interior of a complementary triangle, the geometric structure on F is obviously hyperbolic. At an interior point of a segment, the structure is hyperbolic because we are gluing together two half-planes (each with cone angle π) along geodesic boundaries, giving a hyperbolic surface locally (with cone angle 2π). At the ith switch, we are gluing together two half-planes (each with cone angle π) and the interior of an angle θ_i to produce a total cone angle of $2\pi + \theta_i$. (This is often referred to as a hyperbolic structure with curvature $-\theta_i$ concentrated at the ith switch.)

To correct this effect, we "scallop" the edges of the complementary triangles: that is, we introduce breaks at each switch of τ so that we can control the exterior angle while leaving the edge lengths fixed; see Figure 1.4.6b. Let η_i denote the exterior angle of the region at the ith switch. Actually, *each* of the two sides of a switch corresponds to an exterior angle of a scalloped point on a complementary region, but in our construction, these angles will always be equal. Because the length of an edge is constant under scalloping, we can still glue the scalloped regions together to obtain a singular hyperbolic structure with cone angle $2(\pi - \eta_i) + \theta_i = 2\pi - 2\eta_i + \theta_i$

at the ith switch. For an example, adopt the notation in Figure 1.4.6 for a complementary region. If we only introduce breaks at switches 4 and 5, then the interior angles θ_i change for $i = 1, 2, 3$, not however θ_4 and θ_5. Taking $\eta_i = 0$ simply gives us the unscalloped triangles and the cone angle $2\pi + \theta_i$ of the old structure at the ith switch; when $\eta_i = \theta_i/2$, the structure is non-singular. By changing η_i from zero to $\theta_i/2$, for all i, we can attempt to construct a non-singular hyperbolic surface. However, we have also changed the vertex angles θ_i of the scalloped hyperbolic region. The bulk of the remainder of the proof is devoted to estimates showing that the changes of θ_i are small compared to the changes in the η_i. Then we can change η_i to half the *new* θ_i and use our estimates to show that repeating this process leads to convergence in the values of θ_i and η_i, giving a non-singular hyperbolic structure.

In order to construct the scalloped regions, we shall break them up into two sorts of pieces: triangles and "polygonal slivers"; see Figure 1.4.7. Lemma 1.4.6 below deals with the construction of the slivers and certain estimates which are important in the proof; it is proved by repeated application of Lemma 1.4.5, which covers triangular slivers. Lemma 1.4.7 takes care of the triangular pieces, and in Lemma 1.4.8, these results are combined into the necessary results about the scalloped regions.

FIGURE 1.4.7

The object of these estimates is to choose a value of L, the scale factor of the lengths, sufficiently large to guarantee the convergence discussed above. To this end, we use the following conventions: a statement like "for A sufficiently large" means for A greater than some a priori constant whose value is (implicitly) established in the course of the proof. Similarly, $x = O(y)$ means $|x| \leq K|y|$, where K is a constant which must be independent of the particular value of A (in this example), and $x = o(y)$ means $|x| \leq K|y|$, where K tends to zero as the bound for A (in this example) tends to infinity. Finally, $x \sim y$ means $x = y(1 + o(1))$.

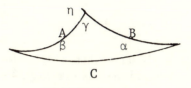

FIGURE 1.4.8

Lemma 1.4.5: *Consider a triangular sliver with edge-lengths and angles as indicated in Figure 1.4.8. Suppose that $0 < \eta < \pi/2$ and A and B are sufficiently large. Then we have the estimates*

(i) $A + B > C > A + B - 1$.

(ii) Suppose that A and η are varied by amounts ΔA and $\Delta \eta$, respectively, where ΔA is bounded and $0 < \eta + \Delta \eta < \pi/2$. If B is held constant, then $\Delta C = O(\Delta \eta) + O(\Delta A)$.

(iii) $\alpha = O(e^{-B} \eta)$, *and* $\beta = O(e^{-A} \eta)$.

(iv) When A and η are varied as in (ii) above, $\Delta \alpha = O(e^{-B} \Delta \eta) + O(e^{-B} \Delta A)$, and $\Delta \beta = O(e^{-A} \Delta \eta) + O(e^{-A} \Delta A)$.

Proof: The proof depends on a fundamental estimate $(*)$ (which will be used again in the proof of Lemma 1.4.7) based on the following hyperbolic law of cosines (see, for instance, [T2;2.6.9]):

$$\cosh C = \cosh A \cosh B - \sinh A \sinh B \cos \gamma$$
$$= \frac{1}{4} e^{A+B} \left[(1 + e^{-2A})(1 + e^{-2B}) - (1 - e^{-2A})(1 - e^{-2B}) \cos \gamma \right]$$
$$= \frac{1}{4} e^{A+B} F(A, B, \gamma) = \frac{1}{4} e^{A+B} F.$$

Now, $\cosh x + \sinh x = e^x$, and $\sinh x = \sqrt{\cosh^2 x - 1}$, so

$$\cosh^{-1} x = \log(x + \sqrt{x^2 - 1}) = \log[x(1 + \sqrt{1 - x^{-2}})],$$

and so

$$C = \log\left[\frac{1}{4}e^{A+B}F(A,B,\gamma)[1 + \sqrt{1 - \frac{e^{-2A-2B}}{\frac{F^2}{16}}}]\right]$$

$$= A + B + \log\frac{F}{2} + \log\frac{1}{2}[1 + \sqrt{1 - \frac{e^{-2A-2B}}{\frac{F^2}{16}}}].$$

If A and B are sufficiently large, then e^{-2A} and e^{-2B} are each $o(1)$, so $F \sim 1 - \cos\gamma$. Using this and the fact that $\log(1 + o(1)) = o(1)$, we obtain the estimate

$$\log\frac{F}{2} = \log\frac{1-\cos\gamma}{2} + o(1).$$

Notice that $\eta = \pi - \gamma$, so $\cos\gamma = -\cos\eta < 0$ and hence $F \geq 1$. Using this, a similar estimate for the second log term gives

$$\log\frac{1}{2}[1 + \sqrt{1 + o(1)}] = \log\frac{1}{2}[1 + 1 + o(1)] = \log(1 + o(1)) = o(1),$$

hence we obtain the fundamental estimate

$$(*) \qquad\qquad C = A + B + \log\frac{1-\cos\gamma}{2} + o(1).$$

To prove (i), as $\cos\gamma < 0$, we find

$$\frac{1-\cos\gamma}{2} \geq \frac{1}{2},$$

so

$$\log\frac{1-\cos\gamma}{2} \geq -\log 2,$$

which implies

$$C > A + B - \log 2 + o(1) > A + B - 1,$$

as desired.

Remark: This implies, for example, that if A and B are sufficiently large, then

$$\frac{\cosh A \cosh B}{\sinh C} \sim \frac{e^{A+B-C}}{2} = O(1).$$

Estimates of this sort will henceforth be used without further comment.

To prove (ii), we use Taylor's theorem to estimate C as a function of A and η. From the formula above for F,

$$\frac{\partial F}{\partial A} = O(1) = \frac{\partial F}{\partial\eta},$$

$$\frac{\partial}{\partial A} \log \frac{F}{2} = \frac{1}{F} \frac{\partial F}{\partial A} = O(1),$$

and

$$\frac{\partial}{\partial \eta} \log \frac{F}{2} = \frac{1}{F} \frac{\partial F}{\partial \eta} = O(1),$$

since $F \geq 1$. Similarly, the partials of the second log term are $O(1)$. Thus,

$$\frac{\partial C}{\partial A} = O(1) = \frac{\partial C}{\partial \eta},$$

and the desired result follows immediately from Taylor's theorem.

To prove (*iii*), we recall the hyperbolic law of sines (see [T2;2.6.16] for instance), which states

$$\frac{\sin \alpha}{\sinh A} = \frac{\sin \gamma}{\sinh C} = \frac{\sin \eta}{\sinh C},$$

and it follows that

$$\alpha = \sin^{-1} \left[\frac{\sinh A \sin \eta}{\sinh C} \right].$$

Furthermore, $\frac{\sinh A}{\sinh C} = O(e^{-B})$, so for B sufficiently large, $\alpha = O(e^{-B}\eta)$, and similarly for β.

To prove (*iv*), notice that the hyperbolic law of sines gives

$$\sin \alpha = \frac{\sinh A}{\sinh C} \sin \eta.$$

Taking $\frac{\partial}{\partial \eta}$ and using (*iii*) to show that $\cos \alpha \sim 1$, we find

$$\begin{aligned}
\cos \alpha \frac{\partial \alpha}{\partial \eta} &= \frac{\sinh A}{\sinh C} \cos \eta - \frac{\sinh A}{\sinh^2 C} \sin \eta \cosh C \frac{\partial C}{\partial \eta} \\
&= \frac{\sinh A}{\sinh C} \cos \eta - \frac{\sinh A}{\sinh C} \sin \eta \frac{\cosh C}{\sinh C} \frac{\partial C}{\partial \eta} \\
&\sim e^{A-C} \cos \eta + e^{A-C} \sin \eta O(1) = O(e^{A-C}).
\end{aligned}$$

To apply Taylor's theorem, we must show that the partials of C satisfy a uniform bound over all permissible values of A and η in a neighborhood of the initial values A_0 and η_0. By hypothesis, A varies in a bounded range, and by Lemma 1.4.5.i, C varies in a bounded range as well. Thus,

$$\frac{\partial \alpha}{\partial \eta} = O(e^{A-C}) = O(e^{A_0 - C_0}) = O(e^{-B_0})$$

in the permitted range of A and η. We abuse notation and henceforth drop the subscript zero on the initial values of the side lengths.

Similarly, taking $\frac{\partial}{\partial A}$, we find

$$\frac{\partial \alpha}{\partial A} \sim e^{A-C} \sin \eta - e^{A-C} \sin \eta \frac{\partial C}{\partial A} = O(e^{A-C}) = O(e^{-B})$$

in the permitted range of A and η. The formulas for $\frac{\partial \beta}{\partial \eta}$ and $\frac{\partial \beta}{\partial A}$ are similar, except that the first term in $\frac{\partial \beta}{\partial A}$ is omitted. The result again follows from Taylor's theorem.   q.e.d.

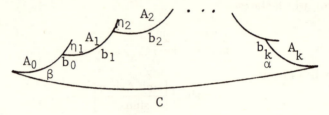

FIGURE 1.4.9

Lemma 1.4.6: *Suppose that angles* $0 \le \eta_i \le \frac{\pi}{4}$ *are given. If L is sufficiently large, there is a polygonal sliver as in Figure 1.4.9, where* $A_i = L\nu(b_i)$, *for* $i = 0, \ldots, k$. C, α, *and* β *are continuous functions of the* η_i, *and we have the estimates*

(i) $\sum_0^k A_i > C > -k + \sum_0^k A_i$,

(ii) *If the* η_i *are varied by an amount* $\Delta\eta_i$ *such that* $0 \le \eta_i + \Delta\eta_i < \frac{\pi}{2}$, *then* $\Delta C = \sum_1^k O(\Delta\eta_i)$,

(iii) $\alpha = \sum_1^k O(e^{-L}\eta_i)$, *and* $\beta = \sum_1^k O(e^{-L}\eta_i)$,

(iv) $\Delta\alpha = \sum_1^k O(e^{-L}\Delta\eta_i)$, *and* $\Delta\beta = \sum_1^k O(e^{-L}\Delta\eta_i)$,

where the various constants may depend on k.

Proof: We proceed by induction on k. Because $\nu > 0$ is an integral tangential measure, $\nu(b) \ge 1$, for each branch b, so making L large will make all the branches long. In the case $k = 1$, the existence of the triangle is immediate, and the estimates are simply those of Lemma 1.4.5. When $k > 1$, we break the sliver up into a sliver with a smaller number of sides and a triangle as in Figure 1.4.10. The exterior angle η of the triangle is simply $\eta_k + \alpha_0$. If L is sufficiently large, then induction hypotheses (iii) and (iv) imply that $\alpha_0 < \frac{\pi}{4}$ and $\alpha_0 + \Delta\alpha_0 < \frac{\pi}{4}$, so that $\eta < \frac{\pi}{2}$ and $\eta + \Delta\eta < \frac{\pi}{2}$, and hence Lemma 1.4.5 applies to the triangle.

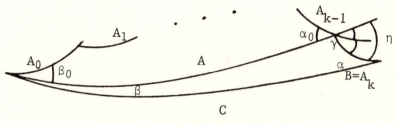

FIGURE 1.4.10

As to the estimates:

(i) follows immediately from Lemma 1.4.5.i and induction hypothesis (i).

For (ii), $\Delta\eta = \Delta\eta_k + \Delta\alpha_0 = \sum_1^k O(\Delta\eta_i)$ by induction hypothesis (iv); so by Lemma 1.4.5.ii, $\Delta C = O(\Delta\eta) + O(\Delta A)$, and estimate (ii) follows immediately from these formulas and induction hypothesis (ii).

For (iii), we have $\eta = \eta_k + \alpha_0 = \sum_1^k O(\eta_i)$ by induction hypothesis (iii); so by Lemma 1.4.5.iii, we have $\alpha = O(e^{-L}\eta) = \sum_1^k O(e^{-L}\eta_i)$. The estimate on β follows by symmetry.

For (iv), we have $\Delta\alpha = O(e^{-L}\Delta\eta) + O(e^{-L}\Delta A)$ by Lemma 1.4.5.iv. Using the formula above for $\Delta\eta$ and induction hypothesis (ii), the estimate on $\Delta\alpha$ follows. Again the estimate on $\Delta\beta$ follows by symmetry.

q.e.d.

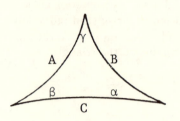

FIGURE 1.4.11

Lemma 1.4.7: *Consider a hyperbolic triangle with edge-lengths and angles as indicated in Figure 1.4.11. If A, B, C, and $A + B - C$ are sufficiently large, then we have the estimates*

(i) $\gamma \sim 2e^{\frac{C-A-B}{2}}$

(ii) If A, B, and C are varied by bounded amounts $\Delta A, \Delta B$, and ΔC, respectively, then

$$\Delta\gamma \sim O(e^{\frac{C-A-B}{2}}\Delta C) + O(e^{\frac{C-A-B}{2}}\Delta A) + O(e^{\frac{C-A-B}{2}}\Delta B)$$

Proof: For (i), the fundamental estimate $(*)$ in the proof of Lemma 1.4.5 says

$$C = A + B + \log\frac{1 - \cos\gamma}{2} + o(1)$$

$$= A + B + \log\sin^2\frac{\gamma}{2} + o(1),$$

so

$$\sin\frac{\gamma}{2} \sim e^{\frac{C-A-B}{2}},$$

and so

$$\frac{\gamma}{2} \sim e^{\frac{C-A-B}{2}},$$

as desired.

For (ii), the hyperbolic law of cosines gives

$$\cos\gamma = \frac{\cosh A \cosh B - \cosh C}{\sinh A \sinh B},$$

so

$$-\sin\gamma\frac{\partial\gamma}{\partial C} = -\frac{\sinh C}{\sinh A \sinh B},$$

and hence

$$\frac{\partial \gamma}{\partial C} \sim \frac{e^{C-A-B}}{e^{\frac{C-A-B}{2}}} = e^{\frac{C-A-B}{2}} .$$

Similarly,

$$-\sin \gamma \frac{\partial \gamma}{\partial A} = \frac{\cosh B}{\sinh B} - \frac{\cosh A}{\sinh B} \ \sinh^{-2} A \ \{\cosh A \cosh B - \cosh C\} ,$$

hence

$$\frac{\partial \gamma}{\partial A} \sim \frac{e^{C-A-B}}{e^{\frac{C-A-B}{2}}} = e^{\frac{C-A-B}{2}} .$$

The symmetric formula follows for $\frac{\partial \gamma}{\partial B}$. Since $A, B,$ and C vary by bounded amounts, $\frac{\partial \gamma}{\partial A} = O(e^{\frac{C_0-A_0-B_0}{2}})$ in the permitted range. The same holds for the other partials, and the result follows from Taylor's theorem.   q.e.d.

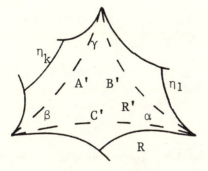

FIGURE 1.4.12

Lemma 1.4.8: *Suppose that angles $0 \le \eta_i < \frac{\pi}{4}$ are given. If L is suffi-ciently large, then there is a scalloped region R with these exterior angles (as in Figure 1.4.12), where the length of a branch b is $L\nu(b)$. $\alpha, \beta,$ and γ are continuous functions of the η_i, and*

(i) $\alpha = O(e^{\frac{-L}{2}})$,

(ii) $\Delta \alpha = \sum O(e^{\frac{-L}{2}} \Delta \eta_i)$,

where identical estimates hold for β and γ.

Proof: By Lemma 1.4.6, we can construct three slivers with long sides of respective lengths A', B', and C', and we let $\nu(A)$, $\nu(B)$, and $\nu(C)$, respectively, denote the total ν-tangential measures of the corresponding frontier edges of the associated complementary sliver. From Lemma 1.4.6.i, $L\nu(C) \geq C', A' > L\nu(A) - \text{constant}$, and $B' > L\nu(B) - \text{constant}$, so

$$A' + B' - C' > L[\nu(A) + \nu(B) - \nu(C)] - \text{constant} \geq L - \text{constant},$$

and similarly for the cyclic permutations of A', B', C'. Thus, if L is sufficiently large, the triangle inequalities hold, and we can construct a hyperbolic triangle R' with side-lengths A', B', C' and opposite interior angles α', β', γ', respectively. Furthermore, for L sufficiently large, Lemma 1.4.7 will apply to this triangle.

The angle α is simply α' plus the angles at the extreme vertices of two of the slivers. By Lemma 1.4.6.iii and Lemma 1.4.7.i, these are both $O(e^{-\frac{L}{2}})$, and estimate (i) follows.

Thus,

$$e^{\frac{C'-A'-B'}{2}} = O(e^{-\frac{L}{2}}),$$

and hence by Lemma 1.4.7.ii

$$\Delta\gamma' = O(e^{-\frac{L}{2}}\Delta A') + O(e^{-\frac{L}{2}}\Delta B') + O(e^{-\frac{L}{2}}\Delta C')$$
$$= \sum O(e^{-\frac{L}{2}}\Delta\eta_i)$$

by Lemmas 1.4.6 and 1.4.7. On the other hand, $\Delta\gamma$ is the sum of $\Delta\gamma'$ and the changes in the extreme angles of the slivers, which are at most $\sum O(e^{-L}\Delta\eta_i) = \sum O(e^{-\frac{L}{2}}\Delta\eta_i)$. Similar estimates hold for α and β, and this completes the proof.   q.e.d.

Proof of Theorem 1.4.4 (conclusion): Choose some $L > L_0$ so that application of Lemma 1.4.8 gives

$$\theta_i < \min(\varepsilon_0, \tfrac{\pi}{8}),$$

$|\Delta\theta_i| \leq \sum |\Delta\eta_j|$, so in particular, $|\Delta\theta_i| \leq |\Delta\eta_i|$.

For $j = 0, 1, 2, \ldots$, we construct a sequence of exterior angles at each switch by taking $\eta_{i0} = 0, \eta_{i1} = \theta_{i0}/2, \ldots, \eta_{ij} = \theta_{i(j-1)}/2, \ldots$, where θ_{ij} is the interior angle of the scalloped region determined by the exterior angles η_{ij} for a fixed j. We will show that the η_{ij} converge to some η_i at least as fast as a geometric series. We have

$$\Delta\eta_{ij} = \eta_{i(j+1)} - \eta_{ij} = \frac{\theta_{ij} - \theta_{i(j-1)}}{2} = \frac{1}{2}\Delta\theta_{i(j-1)}.$$

Using this and the fact that $|\Delta\theta_i| \leq |\Delta\eta_i|$, we see that $|\Delta\theta_{ij}| \leq \frac{|\Delta\theta_{i0}|}{2^j}$, so $\theta_{ij} = \theta_{i0} + \sum_k \Delta\theta_{ik}$ converges to θ_i. Since $\eta_{ij} = \frac{1}{2}\theta_{i(j-1)}$, it follows that η_{ij} converges to $\frac{1}{2}\theta_i$. Since the θ_i are continuous functions of the η_i, Lemma 1.4.8 implies that there is a scalloped region with these particular values of η_i and θ_i, which can be glued together to give a non-singular hyperbolic structure on F. The edges then give a piecewise geodesic imbedding of a CW complex equivalent to τ. Since $L \geq L_0$ and $\theta_i < \varepsilon_0$, the bounds (a) and (b) hold.

Exercise: Complete the proof of Theorem 1.4.4 in the punctured case by providing analyses similar to Lemma 1.4.8 of a "scalloped once-punctured monogon" and a "scalloped once-punctured bigon" (in case $F = F_1^1$); see Figure 1.4.13a. [Hint: Construct such an object by dividing it into slivers and a geodesic once-punctured monogon. A geodesic once-punctured monogon may be constructed by gluing together two congruent hyperbolic right triangles with one ideal vertex; see Figure 1.4.13b. Use the trigonometric formula [T2; 2.16.12] $\sin\alpha \cosh C = 1$.]

This concludes the proof of Theorem 1.4.4   q.e.d.

(a)   (b)

FIGURE 1.4.13

We have thus shown that Condition (ii) implies Condition (iii) in Theorem 1.4.3. It remains to prove the converse, and we suppose that $\tau \subset F$ is a (not necessarily complete) train track so that Condition (iii) holds. Given $\varepsilon_0 > 0$ and $L_0 > 0$, we claim that there is a hyperbolic structure on F and a *piecewise geodesic* CW complex $T \subset F$ so that the following conditions hold.

-T is isomorphic to the underlying CW complex of τ.

-At each vertex of T, the angle of deviation from a straight line is less than ε_0.

-Each edge of T has length exceeding L_0.

Indeed, fix some $\varepsilon > 0$ and $L > 0$, and consider the hyperbolic structure on F guaranteed by Condition (iii). For each branch of τ, choose some bi-infinite trainpath traversing it at least once; this trainpath lifts to a bi-inifinite geodesic curve in hyperbolic space with uniformly small geodesic curvature. As such, this bi-infinite curve lies uniformly close to a unique geodesic. Since τ has only finitely many branches, τ may be "unsmoothed" to a CW complex $T \subset F$, and given any $\varepsilon_0 > 0$ and $L_0 > 0$, ε and L can be chosen so that T satisfies the second and third conditions above. For each branch b of τ, let $\nu(b) \in \mathbb{R}_+$ denote the hyperbolic length of the corresponding edge of T.

We claim that for ε_0 sufficiently small and L_0 sufficiently large, ν is actually a tangential measure on τ as in Condition (ii). Indeed, suppose first that Q is a trigon component of $F - \tau$, and let R be the corresponding component of $F - T$. By the second condition above, R has exactly three vertices with small interior angles (and the remaining interior angles, if any, are near π), and we consider the triangle $R' \subset R$ spanned by these vertices, where the sides of R' have respective lengths A', B', C'; see Figure 1.4.12. Furthermore, let A, B, C, respectively, denote the sums of ν-values of branches in the corresponding frontier edges of Q. Taking L_0 sufficiently large, Lemma 1.4.6.i applies to each component of $R - R'$. If ε_0 is sufficiently small, then the interior angles of R' are small, so we may arrange that $A' + B' - C'$ exceeds $2n$ (by the fundamental estimate $(*)$ in the proof of Lemma 1.4.5), where n is the number of branches of τ. Thus,

$$A + B \geq A' + B' > C' + 2n > C,$$

and similarly for the other triangle inequalities, as desired. The proof that ν satisfies the appropriate strict inequalities on a complementary m-gon, for $m > 3$, follows easily by induction.

Finally, if Q is an m-gon-minus-a-disk component of $F - \tau$, then we may first decompose Q into an $(m+2)$-gon and a monogon-minus-a-disk (where each resulting component has interior angles near 0 and π) by cutting along a geodesic arc in Q with both its endpoints at a vertex on the non-smooth component of the frontier of Q. One establishes the required inequality by arguing separately, as above, in each component.

This completes the proof that Condition (iii) implies Condition (ii) and hence also the proof of Theorem 1.4.3.   q.e.d.

Remark: Notice first that the proofs above that Condition (iii) implies Condition (ii) and Condition (ii) implies Condition (i) did *not* require the hypothesis of genericity on the train track. Genericity of the train track *is* needed however to show that Condition (i) implies Condition (ii). Indeed, consider the non-generic train track $\tau \subset F_2^0$ illustrated in Figure 1.4.3 (and discussed in the example before Corollary 1.4.2). We saw that τ is transversely recurrent, but does not support a tangential measure satisfying strict inequalities in Conditions (1) and (2) in the definition of tangential measure. This also shows that Condition (i) does not imply Condition (iii) without the hypothesis of genericity, as genericity was not needed in the proof that (iii) implies (ii).

To close this section, we describe yet another method of producing a complete generic train track from a birecurrent generic one.

Suppose that $\tau \subset F$ is a (not necessarily birecurrent or generic) train track, R is a component of $F - \tau$, and α is an arc imbedded in R so that $\partial \alpha$ is contained in the frontier of R and $\partial \alpha$ is disjoint from the switches of τ. Provided that no component of $R - \alpha$ is a bigon or trigon, we say that the arc α is *admissible*; see Figure 1.4.14a for an example. One can produce a new track τ' by collapsing a neighborhood of $\partial \alpha$ in the frontier of R to a single arc, as indicated in Figure 1.4.14b; this modification is called a *trivial collapse along* α. Observe that if τ is generic, then the collapse may be performed in such a way that τ' is generic as well.

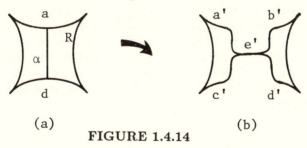

(a)   (b)

FIGURE 1.4.14

We next show that birecurrence is preserved under such a trivial collapse. Adopt the notation indicated in Figure 1.4.14 for the branches in the frontier of R. To any branch f of τ other than a or d, there is a corresponding branch f' of τ'. If μ is a (positive) measure on τ, then there is a corresponding (positive) measure μ' on τ' defined by

$$\mu'(f') = \begin{cases} \mu(a), & \text{if } f' = a' \text{ or } b'; \\ \mu(d), & \text{if } f' = c' \text{ or } d'; \\ \mu(a) + \mu(d), & \text{if } f' = e'; \\ \mu(f), & \text{otherwise.} \end{cases}$$

It follows easily from Proposition 1.3.1 that a trivial collapse along an admissible arc preserves recurrence. Moreover, suppose that τ is transversely recurrent and C is a multiple curve hitting τ efficiently. If some sub-arc α of (some component of) C is admissible, then one sees easily (using parallel copies of components of C) that a trivial collapse of τ along α preserves transverse recurrence, as was asserted.

Proposition 1.4.9: *If $\tau \subset F$ is a birecurrent generic train track, then there is a complete generic train track $\tau' > \tau$ so that τ' arises from τ by a composition of trivial collapses along admissible arcs.*

Proof: We claim that unless τ is complete, there is an admissible sub-arc α of a multiple curve which hits τ efficiently. Performing a trivial collapse along α produces a birecurrent track carrying τ by the remarks above and forms the inductive step in an obvious argument; it therefore remains only to prove the claim.

To this end, since τ is transversely recurrent, there is some multiple curve $C \subset F$ hitting τ efficiently which meets each branch of τ by Corollary 1.3.5. Moreover, since τ is generic, we may sneak up various components of C to furthermore arrange that the following condition holds: if v is a switch of τ giving rise to a non-smooth point in the frontier of a component R of $F - \tau$, then there is an arc among the (closures of) components of $C - \tau$ and a trigon component of $F - (\tau \cup C)$ whose frontier contains both v and this arc. Now, suppose that R is a component of $F - \tau$ so that $\chi(D(R)) < -2$. If no component of $C - \tau$ in R is admissible, then it is an easy matter to surger C along an arc in R to produce another multiple curve meeting τ efficiently which does contain an admissible sub-arc in R; see Figure 1.4.15. Thus, provided $\chi(D(R)) < -2$, we may find a suitable admissible arc in R.

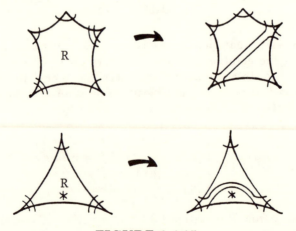

FIGURE 1.4.15

It remains to consider the cases in which a component R of $F - \tau$ is a once-punctured bigon, a fourgon, or a twice-punctured nullgon, and in this last case, every arc among components of $C - \tau$ is admissible. If R is a complementary fourgon, then consider the respective total ν_C-tangential measures $\xi_i, i = 1, \ldots, 4$, of the frontier edges of R enumerated in the order in which they occur in a C^0 traversal of the frontier, where ν_C is the even tangential measure on τ associated to C. It is an exercise to check that provided $\xi_1 + \xi_3 \neq \xi_2 + \xi_4$, there must be an admissible arc among the components of $C - \tau$. If equality holds in the previous equation, then choose a branch b' of τ which occurs exactly once in the frontier of R, and define a new tangential measure ν on τ by

$$\nu(b) = \begin{cases} \nu_C(b) + \varepsilon & \text{if } b = b'; \\ \nu_C(b), & \text{otherwise,} \end{cases}$$

where ε is chosen sufficiently small that ν is actually a tangential measure; such a choice of ε is possible since ν_C satisfies strict inequalities in Conditions (1) and (2) in the definition of tangential measure by construction. As before, we may take a rational approximation to ν, clear denominators, and multiply by two to finally produce an even tangential measure on τ so that the corresponding multiple curve contains an admissible sub-arc in R.

Finally, consider the case of a complementary once-punctured bigon R, and let ξ and ξ', respectively, denote the total ν_C-tangential measures of the frontier edges of R. If $\xi \neq \xi'$, then C contains an admissible sub-arc in R; if equality holds in the previous equation and $F \neq F_1^1$, then a modification as in the previous paragraph produces a suitable admissible arc in R.

Thus, unless τ is complete, we can find an admissible sub-arc of a multiple curve meeting τ efficiently, so the claim and hence the proposition are proved. q.e.d.

§1.5 TRAINPATHS AND TRANSVERSE RECURRENCE

We collect together in this section several results needed later on train-paths in a transversely recurrent train track. Let us fix a surface $F = F_g^s$, where, as usual, $2g - 2 + s > 0$, and if $g = 0$, then $s > 3$. By the Uniformization Theorem (see for instance [A]), there is a complete Riemannian metric of constant Gauss curvature -1 on F, which is called a *hyperbolic structure* on F; thus, the universal cover of F is metrically identified with the upper half-space $\mathbb{H}^2 = \{z = x + y\sqrt{-1} : y > 0\}$ with its hyperbolic metric $ds = \frac{1}{y} \mid dz \mid$. The group of orientation-preserving isometries of $\mathbb{H}^2$, called the *Möbius group*, is naturally isomorphic to the group $PSL(2, \mathbb{R})$ of two-by-two matrices with real entries and determinant one, where we identify each such matrix γ with $-\gamma$; thus, the fundamental group $\pi_1(F)$ of F is identified with a (conjugacy class of) discrete subgroup(s) (of the first kind) in the Möbius group.

Let $S_\infty^1 = \mathbb{R} \cup \{\infty\}$ denote the *circle at infinity*, where $\mathbb{R}$ forms the frontier of $\mathbb{H}^2$, and ∞ is an ideal point which compactifies $\mathbb{R}$ to a circle. Furthermore, let $D \subset S_\infty^1 \times S_\infty^1$ denote the diagonal, and define the *Möbius band* M_∞ *beyond infinity* to be the quotient of $S_\infty^1 \times S_\infty^1 - D$ by the fixed-point-free involution

$$ S_\infty^1 \times S_\infty^1 - D \to S_\infty^1 \times S_\infty^1 - D $$

$$ (x, y) \mapsto (y, x). $$

Unoriented geodesics in $\mathbb{H}^2$ are in one-to-one correspondence with points of M_∞, where the class of a pair $(x, y) \in S_\infty^1 \times S_\infty^1$ corresponds to the geodesic g in $\mathbb{H}^2$ which is asymptotic to x and y. The points x, y are called the *ideal points* of g, and we define $E(g) = \{x, y\} \in M_\infty$. Finally, the action of the group of Möbius transformations on S_∞^1 induces an action of this group on M_∞, and in particular, there are induced (continuous) actions of $\pi_1(F)$ on S_∞^1 and M_∞.

Finally, throughout this section, $\tau \subset F$ will denote a transversely recurrent train track, F is given some fixed hyperbolic structure, and $\tilde{\tau} \subset \mathbb{H}^2$ denotes the full pre-image of τ.

Proposition 1.5.1: *Suppose that τ is a transversely recurrent train track and ρ is a semi-infinite trainpath on $\tilde{\tau}$. There is a unique point $x(\rho) \in S_{\infty}^1$ so that ρ meets every neighborhood of $x(\rho)$ in $\mathbb{H}^2 \cup S_{\infty}^1$. If $\rho' \subset \rho$ is a semi-infinite sub-trainpath on $\tilde{\tau}$, then $x(\rho) = x(\rho')$. Finally, if ρ is a bi-infinite trainpath on $\tilde{\tau}$ and $\rho_1, \rho_2 \subset \rho$ are disjoint semi-infinite sub-trainpaths, then $x(\rho_1) \neq x(\rho_2)$.*

Proof: Since τ has only finitely many branches (say, by Corollary 1.1.3), there is some branch e of τ so that the projection $\bar{\rho}$ of ρ to a trainpath on τ traverses e infinitely many times. By transverse recurrence of τ, there is some *connected* multiple curve C which meets e. We may apply an isotopy of the identity map $F \to F$ to τ as well as to C to arrange that C is a geodesic and consider a lift $\tilde{C}$ of C to $\mathbb{H}^2$. $\tilde{C}$ cuts $\mathbb{H}^2$ into two hyperbolic half-planes, and the pair $E(\tilde{C})$ of ideal points of $\tilde{C}$ cut the circle S_{∞}^1 at infinity into two open intervals. We claim that once ρ enters a component of $\mathbb{H}^2 - \tilde{C}$, then it cannot escape: if ρ meets $\tilde{C}$ in more than one point, then there is an imbedded bigon in $\mathbb{H}^2$ whose closure in $\mathbb{H}^2$ is compact and whose frontier is contained in $\rho \cup \tilde{C}$. It follows that this bigon has its frontier in $\tilde{\tau} \cup \tilde{C}$, and this contradicts that C hits τ efficiently by Proposition 1.3.2. Since $\bar{\rho}$ traverses e infinitely many times, it follows that ρ meets an infinite number $\{\tilde{C}_i : i \geq 1\}$ of lifts of C, where $\tilde{C}_{i+1}$ denotes the next lift of C encountered after $\tilde{C}_i$ in the natural (prescribed) orientation of ρ.

Let $I_i \subset S_{\infty}^1$ denote the closure of the interval with endpoints $E(\tilde{C}_i)$ which contains $E(\tilde{C}_{i+1})$, and suppose (by conjugating by an isometry of $\mathbb{H}^2$, if necessary) that $I_1 \subset \mathbb{R} \subset S_{\infty}^1$ is a finite interval with $I_1 \supset I_2 \supset \ldots$. By compactness of I_1, the intersection of all these intervals is non-empty. By discreteness of the action of $\pi_1(F)$ on $\mathbb{H}^2$, the Euclidean diameters of these intervals tends to zero, so their intersection is in fact a singleton

$$\{x\} = \bigcap_{i \geq 1} I_i,$$

and insofar as ρ meets each neighborhood (in $\mathbb{H}^2 \cup S_{\infty}^1$) of x by construction, this point $x(\rho) = x$ is actually independent of our original choice of multiple curve C.

The assertion in the proposition about a semi-infinite sub-trainpath $\rho' \subset \rho$ holds by construction, and it remains only to prove that the points $x(\rho_1)$ and $x(\rho_2)$ determined respectively by two disjoint semi-infinite sub-trainpaths ρ_1 and ρ_2 of a bi-infinite trainpath ρ are distinct. To this end, if $\tilde{C}$ is any lift to $\mathbb{H}^2$ of C which meets ρ, then $E(\tilde{C})$ decomposes S_{∞}^1 into disjoint open intervals one of which contains $x(\rho_1)$ and the other of which contains $x(\rho_2)$, so $x(\rho_1) \neq x(\rho_2)$. q.e.d.

Remark: From Theorem 1.4.3, it follows immediately that in the appropriate hyperbolic structure, each trainpath on τ is a K-quasi-geodesic, for some K depending only on L_0 and ϵ_0. In other words, each trainpath is K-Hausdorff close to some bi-infinite geodesic in the hyperbolic plane. Proposition 1.5.1 follows immediately.

If ρ is a semi-infinite trainpath on $\tilde{\tau}$, the point $x(\rho)$ is called the *limit point* of ρ. It follows from the previous proposition that a bi-infinite train-path ρ on $\tilde{\tau}$ determines a pair x, y of distinct limit points in S_∞^1. Let $\bar{\rho}$ stand for $\rho \cup \{x, y\}$, and let the class of $\{x, y\}$ in the Möbius band M_∞ be denoted $E(\rho)$. Conversely, such a point of M_∞ essentially uniquely determines the trainpath as we show in Corollary 1.5.3 below.

Proposition 1.5.2: *Suppose that ρ_1 and ρ_2 are each trainpaths on $\tilde{\tau}$, where τ is transversely recurrent. Then there can be no imbedded bigon in $\mathbb{H}^2 \cup S_\infty^1$ whose frontier is contained in $\bar{\rho}_1 \cup \bar{\rho}_2$.*

Proof: We first remark that there can be no imbedded bigon in $\mathbb{H}^2$ with frontier in $\rho_1 \cup \rho_2$ whose closure is compact by Corollary 1.1.2.

Next, we claim that there can be no imbedded bigon B in $\mathbb{H}^2 \cup S_\infty^1$ with frontier in $\bar{\rho}_1 \cup \bar{\rho}_2$ so that one vertex, say p, of B lies in $\mathbb{H}^2$ and the other vertex, say q, lies in S_∞^1. For if so, choose an innermost such bigon, conjugate by an isometry of $\mathbb{H}^2$ (if necessary) so that $q = \infty$, and let e_L, e_R be the branches of $\tilde{\tau}$ which are incident on p and lie in the frontier of B. Index these branches so that when traveling away from p in B, e_L lies to the left and e_R to the right, and let α_L and α_R, respectively, denote the semi-infinite trainpaths corresponding to the frontier edges of B containing e_L and e_R. Define two new trainpaths in $\tilde{\tau}$ starting from p as follows: β_L begins with branch e_L and ever after takes the rightmost possible branch at each switch (like the "righthand rule" for escaping from a maze); similarly, β_R begins with e_R and ever after takes the leftmost branch. See Figure 1.5.1a.

By Corollary 1.1.2 again, the trainpaths $\alpha_L - e_L$, $\alpha_R - e_R$, $\beta_L - e_L$, $\beta_R - e_R$ are all pairwise disjoint (except for the obvious initial points), so β_L and β_R each have ∞ as limit point. Thus, $\beta_L \cup \{\infty\}$ and $\beta_R \cup \{\infty\}$ together bound an imbedded bigon B' in $\mathbb{H}^2 \cup S_\infty^1$ with vertices p and ∞, and by construction, there is no trainpath in $\tilde{\tau} \cap \bar{B}'$ from p into B', where $\bar{B}'$ denotes the closure of B' in $\mathbb{H}^2$. It follows that any trainpath in $\tilde{\tau} \cap \bar{B}'$ which meets B' gives rise to either an imbedded monogon with compact closure in $\mathbb{H}^2$, in contradiction to Corollary 1.1.2, or to an imbedded monogon with ∞ as vertex; see Figure 1.5.1b. Thus, either $\tilde{\tau} \cap B' = \emptyset$, or there is an imbedded monogon in $\mathbb{H}^2 \cup S_\infty^1$ with frontier in $\tilde{\tau}$ and ideal vertex, and these possibilities are not tenable, as we next see.

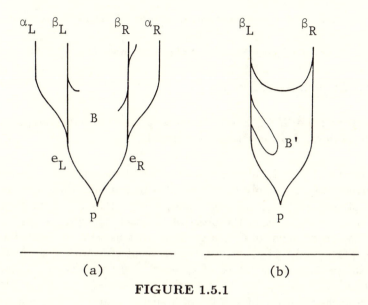

FIGURE 1.5.1

In the first case, any covering translation $\gamma \in \pi_1(F)$ must either map B' off itself or preserve it setwise (as in the last step of the proof of Proposition 1.3.2). However, $\gamma(B') = B'$ implies that $\gamma(p) = p$, so γ is the identity map. It follows that the covering projection is an embedding on B, which is seen to contradict local finiteness of τ.

In the second case, the monogon could project only to a smooth annulus with frontier in τ or to an imbedded monogon component of $F - \tau$, and neither case is possible.

Finally, we prove that there can be no imbedded bigon in $\mathbb{H}^2 \cup S^1_\infty$ with frontier in $\bar{\rho}_1 \cup \bar{\rho}_2$ both of whose vertices lie in S^1_∞. Using the righthand rule as before leads us to either an earlier case or to the possibility that this bigon is imbedded in $\mathbb{H}^2 \cup S^1_\infty - \tilde{\tau}$. Such a bigon could project only to a smooth annulus or to an imbedded bigon component of $F - \tau$, and neither case is possible.   q.e.d.

Corollary 1.5.3: *Suppose that ρ_1 and ρ_2 are each bi-infinite trainpaths on $\tilde{\tau}$, where τ is transversely recurrent. If $E(\rho_1) = E(\rho_2)$, then ρ_1 is equivalent as a trainpath to either ρ_2 or the reverse of ρ_2.*

Proof: It follows from the previous result that if $E(\rho_1) = E(\rho_2)$, then ρ_1 and ρ_2 have exactly the same image in $\mathbb{H}^2$. One easily produces the required equivalence of trainpaths, and the proof is complete.   q.e.d.

For any transversely recurrent train track τ, let us define

$$E(\tau) = \{E(\rho) : \rho \text{ is a bi} - \text{infinite trainpath on } \tilde{\tau}\} \subset \mathbf{M}_\infty.$$

Our main result for this section is

Theorem 1.5.4: *If $\tau \subset F$ is a transversely recurrent train track, then $E(\tau)$ is closed in the Möbius band beyond infinity.*

Proof: Let $b_1, \ldots, b_n$ denote the branches of τ. Since τ is transversely recurrent, we may apply Corollary 1.3.5 to conclude that there is some multiple curve $C \subset F$ hitting τ efficiently so that $C \cap b_i \neq \emptyset$ for each $i = 1, \ldots, n$, and, as in the proof of Proposition 1.5.1, we may assume that each component of C is a geodesic. Let us select for each i a point $p_i \in b_i \cap C$, and let C_i denote the component of C with $p_i \in C_i$. Choose a point $p_0 \in \tau$ and a lift $\tilde{p}_0$ of p_0 to $\mathbb{H}^2$ once and for all. Consider a trainpath $\alpha \colon [-m, m] \to \tilde{\tau}, m \geq 1$, with $p_0 \in \alpha((-\frac{1}{2}, \frac{1}{2}))$. If the first and last branches of α are lifts to $\mathbb{H}^2$ of branches b_i and b_f of τ, respectively, then α determines unique lifts $\tilde{C}_i$ of C_i and $\tilde{C}_f$ of C_f in the natural way; namely, $\tilde{C}_i$ and $\tilde{C}_f$ contain the lifts of p_i and p_f to the first and last branches of α, respectively. Each of $\tilde{C}_i$ and $\tilde{C}_f$ divides $\mathbb{H}^2$ into two hyperbolic half-planes, say $H_i^\pm$ and $H_f^\pm$, respectively, where the superscripts are chosen so that the (oriented) trainpath α passes from H_i^- to H_i^+ and from H_f^- to H_f^+. Let $I_i^\pm$ and $I_f^\pm$, respectively, denote the closed intervals in S_∞^1 which form the ideal boundary of $H_i^\pm$ and $H_f^\pm$, and define

$$J_m(\tilde{p}_0) = \bigcup \; [(I_i^- \times I_f^+) \cup (I_f^+ \times I_i^-)] \subset S_\infty^1 \times S_\infty^1,$$

where the union is over the set of all trainpaths $\alpha \colon [-m, m] \to \tilde{\tau}$ so that $\tilde{p}_0 \in \alpha((-\frac{1}{2}, \frac{1}{2}))$, as above; this is actually a finite union of distinct closed sets, hence $J_m(\tilde{p}_0)$ is itself a closed subset of $S_\infty^1 \times S_\infty^1$, for each $m \geq 1$. Thus, the intersection

$$J(\tilde{p}_0) = \bigcap_{m \geq 1} J_m(\tilde{p}_0)$$

is a closed subset of $S_\infty^1 \times S_\infty^1$. Furthermore, this intersection is nested $J_{m+1}(\tilde{p}_0) \subset J_m(\tilde{p}_0)$, and the limit points of any bi-infinite trainpath on $\tilde{\tau}$ which passes through $\tilde{p}_0$ are contained in $J_m(\tilde{p}_0)$, for all $m \geq 1$. It follows easily that

$$J(\tilde{p}_0) = \{E(\rho) : \rho \text{ is a bi} - \text{infinite trainpath on } \tilde{\tau} \text{ with } \tilde{p}_0 \in \rho\}$$

is a closed subset of $S^1_\infty \times S^1_\infty$, for any $\tilde{p}_0 \in \tilde{\tau}$. In fact, $J(\tilde{p}_0) \subset S^1_\infty \times S^1_\infty - D$ by Corollary 1.5.3 (and $J(\tilde{p}_0)$ is saturated for the projection $S^1_\infty \times S^1_\infty - D \to M_\infty$ by construction), so $J(\tilde{p}_0)$ descends to a closed subset of M_∞ itself.

Now, suppose that $E(\rho_i) = \{x_i, y_i\} \in E(\tau)$, for $i \geq 1$, where $\{\rho_i\}$ is a collection of trainpaths on $\tilde{\tau}$ and $\{x_i, y_i\}$ converges to $\{x, y\} \in S^1_\infty \times S^1_\infty - D$ in the topology of the Möbius band beyond infinity. We may assume that the ρ_i are oriented with initial limit point x_i and terminal limit point y_i in such a way that x_i converges to x and y_i converges to y. If there is a point of $\tilde{\tau}$ through which pass infinitely many ρ_i, then the argument given above applies to show that there is a bi-infinite trainpath on $\tilde{\tau}$ with ideal points x, y, and the theorem follows in this case. We may thus assume that through any point of $\tilde{\tau}$ pass only finitely many ρ_i; let us call this "supposition $(*)$".

We assume for now that $y \notin \{y_i\}$, $x \notin \{x_i\}$, and claim that the ideal points x_i, y_i cannot separate x from y in S^1_∞ for any index i. Suppose, to the contrary, that x_j, y_j do separate x from y, so that infinitely many trainpaths ρ_i must intersect ρ_j. On the other hand, by supposition $(*)$, for any *finite* sub-trainpath $\beta \subset \rho_j$, there can be only finitely many trainpaths among $\{\rho_i\}$ which contain points of β. As in the proof of Proposition 1.5.1 (by discreteness of the action of $\pi_1(F)$ on $\mathbb{H}^2$), since $x \neq y$, there are lifts $\tilde{C}_x$ and $\tilde{C}_y$ to $\mathbb{H}^2$ of components of the multiple curve C above (which hits τ efficiently), so that each of $\tilde{C}_x$ and $\tilde{C}_y$ meets ρ_j (necessarily in a singleton by Proposition 1.3.2), the ideal points $E(\tilde{C}_x)$ separate x_j from x, and the ideal points $E(\tilde{C}_y)$ separate y_j from y. Let $\beta \subset \rho_j$ denote the smallest finite sub-trainpath of ρ_j containing $\rho_j \cap (\tilde{C}_x \cup \tilde{C}_y)$. If ρ_i meets $\rho_j - \beta$, then $E(\tilde{C}_x)$ (and $E(\tilde{C}_y)$, respectively) must separate x_i from x (and y_i from y). This contradicts the convergence of x_i to x and y_i to y, and we have established that no pair x_j, y_j can separate x from y in S^1_∞.

Next, we observe that no two trainpaths on $\tilde{\tau}$ can bound an imbedded bigon in $\mathbb{H}^2$ with compact closure by Proposition 1.5.1 (or Corollary 1.1.2), so, by passing to a subsequence, we may arrange that for any trainpath ρ_j in our sequence, every other trainpath in the sequence lies in the closure of the component of $\mathbb{H}^2 \cup S^1_\infty - \bar{\rho}_j$ which contains $\{x, y\}$.

The pair $\{x, y\}$ decomposes S^1_∞ into two closed intervals I, I', and if we can find $E(\rho_j) \subset I$ and $E(\rho_k) \subset I'$, then argue as follows: choose an arc δ imbedded in $\mathbb{H}^2$ connecting ρ_j and ρ_k, and consider an infinite sequence defined by $z_i \in \rho_i \cap \delta$ if $\rho_i \cap \delta \neq \emptyset$; the sequence $\{z_i\} \subset \delta$ accumulates at some point $z \in \delta$. Since τ is closed in F, it follows that $z \in \tilde{\tau}$, and by discreteness of the action of $\pi_1(F)$ on $\mathbb{H}^2$, the sequence $\{z_i\}$ is eventually constant. Thus, $z \in \tilde{\tau}$ lies in infinitely many trainpaths, in contradiction to supposition $(*)$.

On the other hand, if one of the intervals above, say I, contains no $E(\rho_j)$, then argue as follows. We first claim that there must be some bi-infinite trainpath ρ on $\tilde{\tau}$ (not necessarily in our sequence) with $E(\rho) \subset I$: indeed, we can find an attracting fixed point in I of a hyperbolic motion $\gamma \in \pi_1(F)$ (since $x \neq y$, and such attracting points are dense in S^1_∞); for k sufficiently large, $\rho = \gamma^k(\rho_1)$ is such a trainpath. Of course, for any index i, there can be no imbedded bigon in $\mathbb{H}^2$ with frontier in $\rho \cup \rho_i$ by Proposition 1.5.1. Finally, choose an arc imbedded in $\mathbb{H}^2$ connecting ρ and ρ_j, for some index j, to produce an accumulating sequence as before, in contradiction to supposition $(*)$.

There remains only to consider the case that one of the sequences, say $\{y_i\}$, is constant, so $y_i = y$ and $x_i \neq x$, for all i, and we may conjugate by an isometry of $\mathbb{H}^2$ (if necessary) so that $y = \infty$. Suppose that j and k are indices so that $\{x, x_j\}$ separates x_k from ∞ in S^1_∞. By Proposition 1.5.1, ρ_k lies in the closure of the component of $\mathbb{H}^2 \cup S^1_\infty - \bar{\rho}_j$ which contains x, and one easily produces an accumulating sequence as before, completing the proof.   q.e.d.

Remark: There is a shorter argument using the fact that each bi-infinite trainpath on $\tilde{\tau}$ is K-quasi-geodesic (as in the remark following Proposition 1.5.1) as follows. From supposition $(*)$, it follows that for each compact subset D of $\mathbb{H}^2$, $\rho_i \cap D$ is empty for all but finitely many indices i. If x_i, y_i are sufficiently close to x, y, respectively, then one can choose D appropriately so that if ρ_i misses D, then ρ_i is forced to wander far away from the geodesic g connecting x, y. Indeed, we may choose D to be a disk of large radius centered on a point of g. On the other hand, for each $\epsilon > 0$, there is a $\delta > 0$ such that if $q \subset \mathbb{H}^2 \cup S^1_\infty$ is a K-quasi-geodesic whose ideal points lie within δ of the pair x, y (in the Euclidean metric on $\mathbb{H}^2 \cup S^1_\infty$), then q must stay within $K + \epsilon$ of g (in the hyperbolic metric on $\mathbb{H}^2$). This is a contradiction, because for all sufficiently large i, the ideal points x_i, y_i must be within δ of x, y yet must miss an appropriately chosen D (for instance, a disk D of radius greater than $K + \epsilon$).

As a final note, we add

Proposition 1.5.5: *If $\tau_1, \tau_2 \subset F$ are transversely recurrent train tracks so that $\tau_1 < \tau_2$ (i.e., τ_2 carries τ_1), then $E(\tau_1) \subset E(\tau_2)$.*

Proof: Let $\tilde{\tau}_1$ and $\tilde{\tau}_2$, respectively, denote the full pre-images of τ_1 and τ_2 in $\mathbb{H}^2$. If $\phi : F \to F$ is the supporting map for the carrying $\tau_1 < \tau_2$, then ϕ lifts to a map $\tilde{\phi} : \mathbb{H}^2 \to \mathbb{H}^2$, whose extension to S^1_∞ is the identity map. If ρ_1 is a trainpath on $\tilde{\tau}_1$, then $\tilde{\phi}(\rho_1)$ gives rise to a trainpath ρ_2 on $\tilde{\tau}_2$. If ρ_1 is bi-infinite, then so is ρ_2, and $E(\rho_1) = E(\rho_2)$, so $E(\tau_1) \subset E(\tau_2)$, as was claimed.   q.e.d.

Remark: The converse of this result is false. For instance, suppose that τ_2 is a transversely recurrent track which arises from combing the non-generic track τ_1, so τ_1 is also transversely recurrent by Proposition 1.4.4b. Thus, both $E(\tau_1)$ and $E(\tau_2)$ are defined, and it is not difficult to check that $E(\tau_1) = E(\tau_2)$. On the other hand, τ_1 does not typically carry τ_2 (while τ_2 does carry τ_1), and this gives a counter-example to the converse of the previous result. On the other hand, in the special case that τ_1 is a multiple curve as well as a train track, then the converse is true; see Theorem 1.6.6.

§1.6 LAMINATIONS

A *lamination* L in a surface F is a one-dimensional C^1 foliation of a *closed* subset of F. Thus, L is a collection of C^1 circles and copies of the real line, each of which is called a *leaf*, disjointly imbedded in F, and each point $p \in L$ lies on a unique leaf. If p lies on a leaf ℓ of the latter type, then each "component" (where ℓ is given the topology of $\mathbb{R}$, *not* the topology inherited from $F \supset \ell$) of $\ell - p$ is called a *half-leaf* of ℓ; each half-leaf has infinite length (in any complete Riemannian metric on F) since L foliates a closed subset of F.

As in §1.5, we may choose a hyperbolic structure on F once and for all, so that the universal cover of F is identified with the upper half-space $\mathbb{H}^2$. When each leaf of a lamination $G \subset F$ is a geodesic (for this hyperbolic structure), G is called a *geodesic lamination* in F; the simplest example of a geodesic lamination is a union of disjointly imbedded geodesic curves. General facts about geodesics in a negatively curved manifold imply that no two distinct leaves of a geodesic lamination are parallel (i.e., admit homotopic parametrizations); see [FLP] for the special case we need here.

If g is a geodesic in $\mathbb{H}^2$, then recall that $E(g) \in M_\infty$ denotes the unordered pair of ideal points of g. In general, if $\mathcal{C}$ is a collection of geodesics in $\mathbb{H}^2$, then we define

$$E(\mathcal{C}) = \{E(g) : g \in \mathcal{C}\} \subset M_\infty.$$

For later use and to give an alternate characterization of geodesic laminations (in Corollary 1.6.2 below), we have

Lemma 1.6.1: *If $\mathcal{C}$ is a collection of pairwise disjoint geodesics in $\mathbb{H}^2$, then $\cup\mathcal{C} \subset \mathbb{H}^2$ is closed as a point-set if and only if $E(\mathcal{C})$ is closed in M_∞.*

Proof: Suppose first that $E(\mathcal{C}) \subset M_\infty$ is closed and $p_i \in \cup\mathcal{C}$ accumulates at $p \in \mathbb{H}^2$. Since a geodesic is closed in $\mathbb{H}^2$, we may assume that each p_i lies in a distinct geodesic $g_i \in \mathcal{C}$, where $E(g_i) = \{x_i, y_i\} \in M_\infty$. Compactness of $S^1_\infty \times S^1_\infty$ implies that a subsequence of *ordered* pairs (x_i, y_i) converge, say to $(x, y) \in S^1_\infty \times S^1_\infty$. Since $p_i \in g_i$ converge to $p \in \mathbb{H}^2$, we must have

68

that $x \neq y$, and since $E(\mathcal{C})$ is closed, the geodesic g with ideal points x, y must lie in $\mathcal{C}$. If $p \notin g$, then the hyperbolic half-plane of $\mathbb{H}^2$ determined by g which contains p must contain a neighborhood of p which is disjoint from all but finitely many g_i. This contradicts that p_i converges to p, so $p \in g \subset \cup \mathcal{C}$, as desired.

Conversely, suppose that $\cup \mathcal{C} \subset \mathbb{H}^2$ is closed and $E(g_i) = \{x_i, y_i\} \in E(\mathcal{C})$ converges to $\{x, y\} \in M_\infty$. Choose a geodesic $\hat{g}$ so that $E(\hat{g})$ separates x from y in S_∞^1. Since $\{x_i, y_i\}$ converges to $\{x, y\}$, we must have that $g_i \cap \hat{g} \neq \emptyset$ for i sufficiently large and produce a sequence $p_i = g_i \cap \hat{g} \in \cup \mathcal{C}$. Since $\cup \mathcal{C}$ is closed, p_i accumulates, say at $p \in \mathbb{H}^2$, so there is some $g \in \mathcal{C}$ with $p \in g$. In fact, $E(g) = \{x, y\}$, for if not, then $g_i \cap g \neq \emptyset$ for i sufficiently large, contradicting that the geodesics in $\mathcal{C}$ are pairwise disjoint. Thus, $\{x, y\} \in E(\mathcal{C})$, and the proof is complete. q.e.d.

Remark: Notice that the proof of $E(\mathcal{C})$ closed implies $\cup \mathcal{C}$ closed did not require the geodesics in $\mathcal{C}$ to be pairwise disjoint. To see that this condition is necessary for the converse, simply choose any $\{x, y\} \in M_\infty$ and take $E(\mathcal{C}) = M_\infty - \{\{x, y\}\}$ (so that $\cup \mathcal{C} = \mathbb{H}^2$).

Corollary 1.6.2 *Suppose that G is a geodesic lamination on F, and let $\tilde{G}$ denote the full pre-image of G in $\mathbb{H}^2$. The assignment*

$$G \mapsto E(\tilde{G})$$

establishes a one-to-one correspondence between the collection of all geodesic laminations in F and the collection of all closed, $\pi_1(F)$-invariant subsets $X \subset M_\infty$ with the property that no member of X separates the elements of any other member of X in S_∞^1.

Proof: This follows directly from the definitions, the previous result, and the observation that $G \subset F$ is closed if and only if $\tilde{G} \subset \mathbb{H}^2$ is closed (since $\pi_1(F)$ acts discontinuously on $\mathbb{H}^2$). q.e.d.

Remark: The previous result gives a natural topology on the set of all geodesic laminations in F; namely, as a subspace of the Hausdorff topology on closed subsets of M_∞.

In his analysis of surface diffeomorphisms, Jakob Nielsen (see, for instance, [N] for an introductory account and further references) was led to consider certain regions in the surface canonically associated with iterates of lifts to $\mathbb{H}^2 \cup S_\infty^1$ of a diffeomorphism of the surface. In effect, these regions correspond to the components of $F - G$ for $G \subset F$ a geodesic lamination. In particular, our next result is essentially due to Nielsen.

Proposition 1.6.3: *If $G \subset F$ is a geodesic lamination, then $F - G$ has only finitely many components, and G has zero area (where the area element arises from the hyperbolic structure). In particular, each component of $F - G$ is an imbedded sub-surface of finite area whose frontier consists of finitely many elements of G; furthermore, the double of such a region along these geodesics is a surface with negative Euler characteristic.*

Proof: Suppose that R is a component of $F - G$ and $x \in R$. Since G is closed, R contains some neighborhood of x, so R is an imbedded sub-surface and hence has finite area. Let $\tilde{G}$ denote the full pre-image of G in $\mathbb{H}^2$, let $\tilde{x} \in \mathbb{H}^2$ be some lift of x, and consider the component $\tilde{R}$ of $\mathbb{H}^2 - \tilde{G}$ which contains $\tilde{x}$. The frontier of $\tilde{R}$ consists of a collection of geodesics in $\tilde{G}$ by the previous lemma, so the frontier of R likewise consists of a collection of elements of G, and indeed this latter collection is finite since R has finite area. Thus, the double D of R along these geodesics is a surface without boundary which supports a hyperbolic structure (inherited from that on R), so D has negative Euler characteristic. Furthermore, the area of D is at least 2π, so the area of R is at least π, and it follows that $F - G$ has only finitely many components.

It remains to show that G has zero area. In light of the characterization of components of $F - G$ proved above, one can add a *finite* collection of geodesics in F to G to produce a geodesic lamination G' in F so that each component of $F - G'$ is an (open) ideal triangle. Since $G' \supset G$, it suffices to show that G' has zero area. The proof is analogous to that of Proposition 1.1.1 and runs as follows.

Let $T \subset F$ be a triangle complementary to G' in F, and extend G' to a foliation of T with one three-pronged singular point $V \in T$ as indicated in Figure 1.6.1. Performing this extension for each such region T, we arrive at a foliation $\mathcal{F}$ of F which is C^1 (after some smoothing) except for a finite collection, say $\{V_i\}_{i=1}^{M}$, of three-pronged singularities, one in each component of $F - G'$.

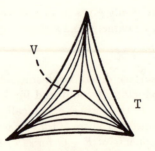

FIGURE 1.6.1

Since the hyperbolic area of an ideal triangle is π, it follows that the area of $F - G'$ is $M\pi$. Furthermore, we claim that

(∗). $$M = -2\chi(F)$$

Assuming this for a moment, the Gauss-Bonnet Theorem implies that the area of F is

$$-2\pi\chi(F) = -2\pi\frac{M}{2} = M\pi,$$

so F and $F - G'$ have the same area, and the proposition follows.

It remains only to prove (∗), and we first suppose that M is even (as must be the case if (∗) is valid). Consider the double cover $\tilde{F}_0$ of F branched over $\{V_i\}_{i=1}^M$; $\mathcal{F}$ lifts to a foliation $\tilde{\mathcal{F}}_0$ on $\tilde{F}_0$ which is regular except for M six-pronged singularities. By the Riemann-Hurwitz formula

$$\chi(\tilde{F}_0) = 2\chi(F) - M.$$

Furthermore, since M is even, we may build a vectorfield tangent to $\tilde{\mathcal{F}}_0$ with exactly M zeroes, each of order -2, so by the Poincaré-Hopf Theorem

$$\chi(\tilde{F}_0) = -2M,$$

and (∗) follows in this case. If M is odd, choose a regular point V_0 of the foliation $\mathcal{F}$, and consider instead the double cover $\tilde{F}_1$ of F branched over $\{V_i\}_{i=0}^M$, so that $\mathcal{F}$ lifts to a foliation $\tilde{\mathcal{F}}_1$ of $\tilde{F}_1$ which is regular except for M six-pronged singularities and one four-pronged singularity. In this case, the Riemann-Hurwitz formula gives

$$\chi(\tilde{F}_1) = 2\chi(F) - (M + 1).$$

Since $M+1$ is even, we may build a vectorfield tangent to $\tilde{\mathcal{F}}_1$ with exactly M zeroes of order -2 and one zero of order -1, and the Poincaré-Hopf Theorem gives

$$\chi(\tilde{F}_1) = -2M - 1,$$

so (∗) follows in this case as well, completing the proof.   q.e.d.

Suppose that $G \subset F$ is a geodesic lamination with full pre-image $\tilde{G} \subset \mathbb{IH}^2$. A *frontier leaf* of $\tilde{G}$ is a leaf lying in the frontier of a component of $\mathbb{IH}^2 - \tilde{G}$, and a *frontier leaf* of G is the image under covering projection of a frontier leaf of $\tilde{G}$. More intrinsically, g is a frontier leaf of G if there is a path $p : [0,1] \to F$ so that $[0,1) \subset p^{-1}(F - G)$ and $p(1) \in g$.

A geodesic lamination G in F is said to have *compact support* if there is a regular neighborhood of Δ in $\hat{F}$ (where, we recall, Δ is the collection

of distinguished points in $\hat{F}$ giving rise to the punctures of $F = \hat{F} - \Delta$)
which is disjoint from G; in particular, every geodesic lamination in a closed
surface $F = F_g^0$ has compact support. We define $\mathcal{U} \subset F$ to be the union of
all the horocycles of length less than one about all the punctures of F, so,
in particular, $\mathcal{U} = \emptyset$ if F is closed.

Proposition 1.6.4: $\mathcal{U} \cup \Delta$ *is a regular neighborhood of* Δ *in* $\hat{F}$, *and a*
geodesic lamination G *in* F *has compact support if and only if* $G \cap \mathcal{U} = \emptyset$.

Proof: If $\Delta = \emptyset$, then there is nothing to prove, and otherwise, for each
$p \in \Delta$, consider the collection $\mathcal{U}_p$ of horocycles of length less than one about
p.

We first claim that $\mathcal{U}_p$ is a deleted regular neighborhood of p in F,
for each $p \in \Delta$. To see this, fix attention on some such p and conjugate
by an isometry of $\mathbb{H}^2$ so that the primitive (in $\pi_1(F)$) parabolic covering
transformation $\gamma = \left(\begin{smallmatrix} 1 & 1 \\ 0 & 1 \end{smallmatrix}\right) \in PSL(2, \mathbb{R})$ corresponds to p; thus, $H = \{z =$
$x + y\sqrt{-1} \in \mathbb{H}^2 : y > 1\}$ is a component of the pre-image of $\mathcal{U}_p$ in $\mathbb{H}^2$. Let
$\gamma_1 = \left(\begin{smallmatrix} a & b \\ c & d \end{smallmatrix}\right) \in PSL(2, \mathbb{R})$ denote an arbitrary element of $\pi_1(F)$ which is not
a power of γ: we must show that $H \cap \gamma_1(H) = \emptyset$. Now, $| c | \geq 1$, for if not,
then the inductive definition $\gamma_{k+1} = \gamma_k \circ \gamma \circ \gamma_k^{-1}$ produces a sequence of
distinct covering transformations with γ_k converging to γ, and this violates
discreteness of $\pi_1(F)$. The image $\gamma_1(H)$ is the interior of a Euclidean disk
which is tangent to $\mathbb{R} \subset S_\infty^1$ at the point $\frac{a}{c}$, and a short computation
shows that the diameter of this disk is bounded above by $| c |^{-2} \leq 1$. Thus,
$H \cap \gamma_1(H) = \emptyset$, and since γ_1 was arbitrary, our claim is proved.

If p' is another element of Δ, then we next assert that $\mathcal{U}_p \cap \mathcal{U}_{p'} = \emptyset$. Indeed, keeping the normalizations and notations of the previous paragraph,
consider a component H' of the full pre-image of $\mathcal{U}_{p'}$, so H' is the interior
of some Euclidean disk tangent to $\mathbb{R} \subset S_\infty^1$. If $H \cap H' \neq \emptyset$, then this disk
must have diameter exceeding unity, so $H' \cap \gamma(H') \neq \emptyset$, and this contradicts
that $\mathcal{U}_{p'}$ is a deleted neighborhood of p' in $\hat{F}$.

It follows that $\mathcal{U} \cup \Delta$ is indeed a regular neighborhood of Δ in $\hat{F}$, and it
remains to show only that if a geodesic lamination G meets $\mathcal{U}$, then G does
not have compact support. Suppose that a leaf g of G meets a component,
say $\mathcal{U}_p$, of $\mathcal{U}$. Again we adopt the normalizations and notations of the earlier
paragraph, and consider a lift $\tilde{g}$ of g to $\mathbb{H}^2$ so that $\tilde{g} \cap H \neq \emptyset$. If g does
not meet every neighborhood of p, then $\tilde{g}$ is a Euclidean semi-circle which
is perpendicular to $\mathbb{R} \subset S_\infty^1$, and since $\tilde{g}$ meets H, this semi-circle must
have radius exceeding unity. It follows that $\tilde{g} \cap \gamma(\tilde{g}) \neq \emptyset$, contradicting that
g is simple.   q.e.d.

Remark: The proof above that $\mathcal{U}_p$ is imbedded is borrowed from [A] and included for completeness.

We say that a lamination (or geodesic lamination) L in F is *carried by* a train track $\tau \subset F$ if there is a *supporting map* $\phi : F \to F$ satisfying the same three conditions as in the definition of carrying in §1.2 (where L is substituted for the multiple curve C) and write $\tau > L$, as before, if τ carries L.

Theorem 1.6.5: *Every geodesic lamination G in F with compact support is carried by some transversely recurrent train track $\tau \subset F$.*

Proof: Arguing as in the proof of Proposition 1.6.3 above, we may add a finite collection of geodesics in F to G to produce a geodesic lamination G', whose full pre-image in $\mathbb{H}^2$ is denoted $\tilde{G}'$, so that each component of $\mathbb{H}^2 - \tilde{G}'$ is either an open ideal triangle or perhaps the interior of an infinite-sided ideal polygon whose vertices comprise the orbit of some point of S^1_∞ under a cyclic parabolic subgroup of $\pi_1(F)$; each component of the latter type projects to a once-punctured region in F whose frontier consists of a single geodesic in G', and as before, each component of the former type imbeds in F. Such an extension is possible since G has compact support, so $G \subset F - \mathcal{U}$ by Proposition 1.6.4, and the resulting geodesic lamination G' has compact support as well

The closed ε-neighborhood $\tilde{N}_\varepsilon$ of $\tilde{G}'$ in $\mathbb{H}^2$ has piecewise C^1 boundary ∂N_ε with one cusp (i.e., non-smooth point) for each pair $(\tilde{R}, v)$, where $\tilde{R}$ is a component of $\mathbb{H}^2 - \tilde{G}'$ and v is an ideal vertex of $\tilde{R}$. $\tilde{N}_\varepsilon$ projects to the ε-neighborhood N_ε of G' in F, and the boundary ∂N_ε is piecewise C^1 with only a *finite* number of cusps; indeed, the number of cusps is given by a constant K which depends only on the topological type of F; namely, $K = -6\chi(F) - 2s$ as in the proof of Proposition 1.6.3 (where, as usual, s denotes the number of punctures of F).

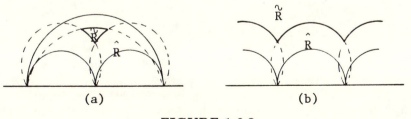

FIGURE 1.6.2

Since $F - G'$ has a finite number of finite-sided components by Proposition 1.6.3, G' has only finitely many (disjoint) frontier leaves, and we may choose $\varepsilon > 0$ with the following property: if $\tilde{g}$ and $\tilde{g}'$ are frontier leaves of $\tilde{G}'$ arising from the same component of $\mathbb{H}^2 - \tilde{G}'$, then the hyperbolic distance from $\tilde{g}$ to $\tilde{g}'$ is less than 3ε if and only if $E(\tilde{g}) \cap E(\tilde{g}') \neq \emptyset$. Thus, for any component $\tilde{R}$ of $\mathbb{H}^2 - \tilde{N}_\varepsilon$, there is a component $\hat{R}$ of $\mathbb{H}^2 - \tilde{G}'$ so that each frontier edge of $\tilde{R}$ lies in an ε-hypercycle of the corresponding frontier geodesic of $\hat{R}$; see Figure 1.6.2a and b. (Recall that an "ε-hypercycle" h_ε about a hyperbolic geodesic $\tilde{g}$ is a component of the frontier of the metric ε-neighborhood of $\tilde{g}$. The two key fact for us is: the component of $\mathbb{H}^2 - h_\varepsilon$ containing $\tilde{g}$ is convex.)

Next, we claim that there is some $\delta > 0$ so that given any two points $p, q \in G'$ that are within δ of one another, the tangent lines to the leaves of G' through p and q are close. To see this, suppose that $\tilde{g}_1, \tilde{g}_2$ are distinct geodesics in $\tilde{G}'$ and $\tilde{p}_i \in \tilde{g}_i, i = 1, 2$. Let $E(\tilde{g}_1) = \{x, y\}$ and construct the geodesic $\tilde{g}_x$ (and $\tilde{g}_y$ respectively), which passes through $\tilde{p}_2$ and has x (and y respectively) as one of its ideal points. Since $\tilde{g}_1$ and $\tilde{g}_2$ are disjoint, it follows that $E(\tilde{g}_2)$ separates the members of $E(\tilde{g}_x)$ on S^1_∞; similarly, $E(\tilde{g}_2)$ separates the members of $E(\tilde{g}_y)$. As $\tilde{p}_1$ tends to $\tilde{p}_2$, each of $\tilde{g}_x$ and $\tilde{g}_y$ approach $\tilde{g}_1$, so $\tilde{g}_2$ approaches $\tilde{g}_1$, and our claim finally follows from compactness of $F - \mathcal{U}$ since $G' \subset F - \mathcal{U}$ by Proposition 1.6.4.

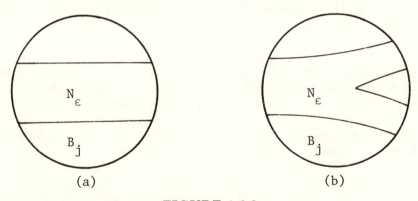

FIGURE 1.6.3

Now, since G' has zero area, the area of $F - N_\varepsilon$ approaches that of F as ε tends to zero. Therefore, N_ε contains no disk of radius δ for ε sufficiently small. It is thus possible to cover N_ε with a finite number of balls $\{B_j\}$ of radius δ, each of which meets N_ε in one of the two ways illustrated in Figure 1.6.3a and b, respectively. In the former case, B_j contains two smooth segments of ∂N_ε, and we regard $B_j \cap N_\varepsilon$ as a nearly "horizontal" strip. By our choice of δ, the intersection of each leaf of G' with B_j is also nearly horizontal and is therefore transverse to the perpendicular "vertical" direction in B_j; let us foliate $B_j \cap N_\varepsilon$ by vertical arcs, so the foliation constructed is transverse to G' in B_j. In the latter case, $B_j \cap N_\varepsilon$ contains one or more cusps of ∂N_ε. Several of the B_j may be required to cover the region around these cusps, but at most $K+1$ are necessary. Again by our choice of δ, all the leaves of G' in such a region have close tangent directions, which we regard as nearly "horizontal", and we may construct a foliation of N_ε in this region by "vertical" arcs which are transverse to G' inside the region.

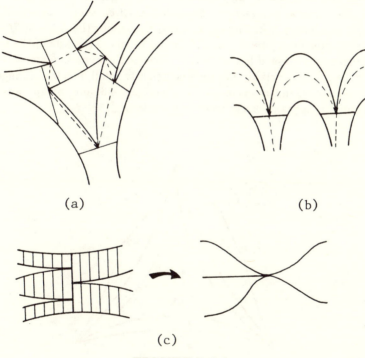

(a)   (b)

(c)

FIGURE 1.6.4

It is an easy matter to deform and combine the foliations described above to produce a C^1 foliation $\mathcal{F}^\perp$ of N_ε whose leaves are everywhere transverse to the leaves of G'. A leaf of $\mathcal{F}^\perp$ is said to be "singular" if it passes through a cusp of ∂N_ε, and we may suppose that each singular leaf of $\mathcal{F}^\perp$ is a geodesic segment. By construction, the singular leaves of $\mathcal{F}^\perp$ decompose the interior $\overset{\circ}{N}_\varepsilon$ into a collection $\{r_i\}_{i=1}^n$ of convex "rectangles"; see Figure 1.6.4a and b, where we depict a lift to $\mathbb{H}^2$ (and the dashed lines are yet to be explained). Collapsing each leaf of $\mathcal{F}^\perp$ to a point describes a map $\phi: F \to F$ which is homotopic to the identity; let τ_ε denote the quotient of N_ε by this collapse and regard τ_ε as a subset of F in the natural way. We claim that τ_ε is a train track. Each rectangle r_i gives rise to a corresponding branch b_i of τ_ε, $i = 1, \ldots, n$. Switches of τ_ε arise from singular leaves of $\mathcal{F}^\perp$ as in Figure 1.6.4c, and there is a natural notion of incoming and outgoing at each resulting switch. Finally, the complementary regions of τ_ε are either trigons or once-punctured monogons by construction, so $\tau_\varepsilon \subset F$ is in fact a train track. Furthermore, Condition (3) in the definition of carrying follows from the transversality of $\mathcal{F}^\perp$ and G', so $G' < \tau_\varepsilon$.

It remains to show that τ_ε is transversely recurrent if ε is chosen to be sufficiently small, and we may assume that τ_ε is generic (since τ_ε is generic for all but countably many ε). To each branch b_i, there is a well-defined geodesic segment α_i with endpoints $\partial \alpha_i \subset \partial r_i$ among the cusps of ∂N_ε, and $\alpha_i \subset r_i$ by convexity; see Figure 1.6.4a and b, where the lifts of the arcs α_i are indicated by dashed lines. Let $\nu(b_i)$ denote the hyperbolic length of α_i, for $i = 1, \ldots, n$. We claim that for ε sufficiently small, $\nu > 0$ is a tangential measure on τ_ε as in Condition (ii) of Theorem 1.4.3, and, by application of this result, τ_ε is transversely recurrent, as desired.

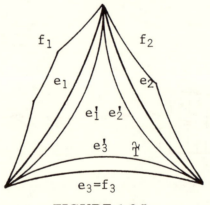

FIGURE 1.6.5

To verify Condition (ii), consider a trigon $\tilde{T}$ among the components of $\mathbb{H}^2 - \tilde{N}_\varepsilon$, and let T denote the corresponding trigon among components of $F - N_\varepsilon$. The arcs $\{\alpha_i\}$ which lie in rectangles adjacent to T lift to $\mathbb{H}^2$ and combine in the natural way to give three disjointly imbedded piecewise geodesic arcs f_j, $j = 1, 2, 3$, one such arc connecting each pair of vertices of $\tilde{T}$. On the other hand, each corresponding frontier edge e'_j of $\tilde{T}$ is an ε-hypercyclic arc (about the corresponding ideal edge of $\hat{T} \supset \tilde{T}$), and

$$e'_j \cap (f_1 \cup f_2 \cup f_3) = \partial e'_j \text{ for } j = 1, 2, 3$$

by convexity of rectangles. Consider the convex hull $C(\tilde{T})$ of the vertices of $\tilde{T}$; $C(\tilde{T})$ is a triangle, say with frontier edges e_j corresponding to e'_j, respectively. By convexity (of ε-hypercyclic neighborhoods), we must have $C(\tilde{T}) \supset \tilde{T}$, and each $f_j \cup e_j$ bounds a (perhaps degenerate) geodesic sliver (in the sense of §1.4); the length of e_j tends uniformly to infinity (and the interior angles of $C(\tilde{T})$ tend uniformly to zero) as ε goes to zero by construction, for $j = 1, 2, 3$; see Figure 1.6.5.

Consider the geodesic sliver corresponding to the edge e_j of $C(\tilde{T})$ and suppose that f_j consists of M geodesic segments; the length of f_j is evidently bounded above by the length of e_j plus $2M\varepsilon$. Since M is bounded (in terms only of the topological type of F), we may choose ε sufficiently small that the length of f_j differs from the length of e_j by at most unity, and we may guarantee that this estimate holds for each geodesic sliver.

Arguing as in the last step of the proof of Theorem 1.4.3, let A, B, and C, respectively, denote the lengths of the frontier edges of $C(\tilde{T})$. By the fundamental estimate $(*)$ in the proof of Lemma 1.4.5, we may take ε so small that $A + B > C + 1$ since $\tilde{T}$ has long edges and small interior angles. Finally, if A', B', and C', respectively, denote the corresponding lengths of the piecewise geodesic segments in the frontier of the sliver, then

$$A' + B' \geq A + B > C + 1 > C'.$$

Similarly, the other strict triangle inequalities hold among A', B', and C', so ν satisfies the strict inequalities in Condition (1) in the definition of tangential measure. Since complementary regions of τ are either trigons or once-punctured monogons, it follows that Condition (ii) in Theorem 1.4.3 holds, and our theorem is proved.   q.e.d.

Remark: The first part of this proof is borrowed from [Cb]. [Tg] proves this result by passing to the universal cover and flowing along horocyclic arcs to define the collapse of G onto a train track. Thurston gives a transverse Lipshitz structure to an arbitrary measured geodesic lamination G in F.

We close this section with a result which completely characterizes when a transversely recurrent train track τ carries a geodesic lamination G with compact support in terms of containment of the corresponding sets $E(\tilde{G})$ and $E(\tau)$.

Theorem 1.6.6: *Suppose that $\tau \subset F$ is a transversely recurrent train track and G is a geodesic lamination in F of compact support with full pre-image $\tilde{G} \subset \mathbb{H}^2$. Then $E(\tilde{G}) \subset E(\tau)$ if and only if $G < \tau$.*

Proof: Let $\tilde{\tau}$ denote the full pre-image of τ in $\mathbb{H}^2$. As in the proof of Proposition 1.5.5, if $G < \tau$, say with supporting map ϕ, then ϕ lifts to a map

$$\tilde{\phi}: \mathbb{H}^2 \cup S_\infty^1 \to \mathbb{H}^2 \cup S_\infty^1$$

which fixes S_∞^1 pointwise. If $\tilde{g} \subset \mathbb{H}^2$ is a lift of some leaf of G, then $\tilde{\phi}(\tilde{g})$ gives rise to a bi-infinite trainpath $\tilde{\rho}$ on $\tilde{\tau}$. Since $\tilde{\phi}$ fixes S_∞^1 pointwise, $E(\tilde{\rho}) = E(\tilde{g})$, so $E(\tilde{G}) \subset E(\tau)$, as desired.

Conversely, by Corollary 1.3.5, there is some multiple curve $C \subset F$ hitting τ efficiently which meets every branch of τ. We may assume (by simply deleting components) that every component of C actually meets τ. Furthermore, by applying an isotopy of the identity map of F to τ as well as to C, we may assume that each component of C is a geodesic.

Under these conditions, we first claim that no component of C can be a leaf of G. For, if $g \subset C$ is a simple closed curve leaf of G and $\tilde{g}$ is some lift of g to $\mathbb{H}^2$, then there must be a trainpath $\tilde{\rho}$ on $\tilde{\tau}$ with $E(\tilde{\rho}) = E(\tilde{g})$ since $E(\tilde{G}) \subset E(\tau)$ by hypothesis. Since C meets every branch of τ, there is a lift $\tilde{c}$ of some component of C which meets $\tilde{\rho}$, and since C hits τ efficiently, $E(\tilde{c})$ must separate the members of $E(\tilde{\rho}) = E(\tilde{g})$ on S_∞^1. It follows that $\tilde{g} \cap \tilde{c} \neq \emptyset$, which contradicts that C is a multiple curve. Our claim is therefore proved, and it follows that C and G meet transversely wherever they intersect; of course, C also meets τ transversely (by definition of hitting efficiently).

Now, fix a component c of C, suppose that $z \in c \cap g$, for some leaf g of G, and choose a lift $\tilde{z}$ of z to $\mathbb{H}^2$. There are well-defined lifts $\tilde{g}$ and $\tilde{c}$, respectively, of g and c to $\mathbb{H}^2$ so that $\tilde{z} \in \tilde{c} \cap \tilde{g}$. Since $E(\tilde{G}) \subset E(\tau)$, there is a trainpath $\tilde{\rho}$ on $\tilde{\tau}$ with $E(\tilde{\rho}) = E(\tilde{g})$, and indeed there is a unique such trainpath (up to reversal) by Corollary 1.5.3. Since c hits τ efficiently, $\tilde{c} \cap \tilde{\rho}$ is a singleton, say $\{\tilde{z}'\}$, and $\tilde{z}' \in \mathbb{H}^2$, in turn, projects to a point $z' \in \tau \cap c$. The assignment $z \mapsto z'$ is clearly independent of the choice of lift $\tilde{z}$ of z and gives a well-defined map

$$\beta: G \cap c \to \tau \cap c.$$

We claim that the assignment

$$\tilde{G} \cap \tilde{c} \to \tilde{\tau} \cap \tilde{c}$$

$$\tilde{z} \mapsto \tilde{z}'$$

above has the following property: if the subarc A of $\tilde{c}$ with endpoints $\tilde{z}_1, \tilde{z}_2 \in \tilde{G} \cap \tilde{c}$ imbeds in F, then the subarc of $\tilde{c}$ with endpoints $\tilde{z}_1', \tilde{z}_2' \in \tilde{\tau} \cap \tilde{c}$ does as well. For suppose not, and choose the subscripts so that there is some point $\tilde{z}_1''$ between $\tilde{z}_1'$ and $\tilde{z}_2'$ on $\tilde{c}$ which projects in F to the same point as $\tilde{z}_1'$. For $i = 1, 2$, let $\tilde{g}_i$ be the lift to $\mathbb{H}^2$ of the leaf of G so that $\tilde{c} \cap \tilde{g}_i = \{\tilde{z}_i\}$, and let $\tilde{\rho}_i$ be the trainpath on $\tilde{\tau}$ as above where $E(\tilde{g}_i) = E(\tilde{\rho}_i)$ and $\tilde{c} \cap \tilde{\rho}_i = \{\tilde{z}_i'\}$; see Figure 1.6.6a. Of course, $E(\tilde{\rho}_1) = E(\tilde{g}_1)$ cannot separate the members of $E(\tilde{\rho}_2) = E(\tilde{g}_2)$ on S_∞^1 since the leaves of G are disjointly imbedded.

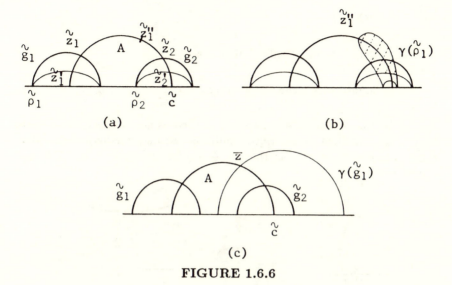

(a) (b)

(c)

FIGURE 1.6.6

Now, there is a hyperbolic covering translation $\gamma \in \pi_1(F)$ leaving $\tilde{c}$ invariant with $\gamma(\tilde{z}_1') = \tilde{z}_1''$, and $\gamma(\tilde{\rho}_1)$ is a trainpath on $\tilde{\tau}$. As before, $E(\gamma(\tilde{g}_1)) = E(\gamma(\tilde{\rho}_1))$ cannot separate the members of $E(\tilde{g}_1) = E(\tilde{\rho}_1)$ or $E(\tilde{g}_2) = E(\tilde{\rho}_2)$ on S^1_∞, and we investigate the various possibilities. If $E(\tilde{\rho}_2)$ separates $E(\gamma(\tilde{\rho}_1))$ from $E(\tilde{\rho}_1)$, then we are led to an imbedded bigon in $\mathbb{H}^2 \cup S^1_\infty$ (perhaps with ideal vertex) whose frontier is contained in $\tilde{\tau}$, contradicting Proposition 1.5.2; see Figure 1.6.6b. Similarly, $E(\tilde{\rho}_1)$ cannot separate $E(\gamma(\tilde{\rho}_1))$ from $E(\tilde{\rho}_2)$. We are thus led to the conclusion that $E(\gamma(\tilde{\rho}_1))$ separates $E(\tilde{\rho}_1)$ from $E(\tilde{\rho}_2)$, so $E(\gamma(\tilde{g}_1))$ separates $E(\tilde{g}_1)$ from $E(\tilde{g}_2)$ on S^1_∞. Thus, $\gamma(\tilde{g}_1) \cap \tilde{c}$ is a singleton, say $\{\tilde{z}\}$, and $\tilde{z} \in A$ since leaves of G are disjointly imbedded; see Figure 1.6.6c. Of course, $\gamma(\tilde{z}_1) = \tilde{z}$, and this finally violates the hypothesis that A imbeds in F.

Our claim is therefore proved, and we next derive some immediate consequences. It follows easily that if $z_j \in c \cap G, j = 1, 2, 3$, occur in the order z_1, z_2, z_3 along c (in some fixed orientation of c), then $\beta(z_j) \in c \cap \tau, j = 1, 2, 3$, must occur in the order $\beta(z_1), \beta(z_2), \beta(z_3)$. (Of course, the $\beta(z_j)$ need not all be distinct.) Consequently, if $z_1 \neq z_2 \in c \cap G$ satisfy $\beta(z_1) = \beta(z_2)$ and a_1, a_2 are the components of $c - \{z_1, z_2\}$, then the following holds for at least one element $a \in \{a_1, a_2\}$: for any $z \in a \cap \tau$, we must have $\beta(z) = \beta(z_1) = \beta(z_2)$.

Now, fix a component c of C and enumerate $c \cap \tau = \{x_j\}_{j=1}^J$ in the order determined by an orientation of c. It follows easily from the remarks above and the fact that $G \cap c$ is closed (since G is a lamination) that there is a collection $\{a_j\}_{j=1}^J$ of disjoint, non-empty, open sub-intervals of c so that the following conditions hold.

$-c \cap G \subset \cup_{j=1}^{J} a_j.$

$-u \in a_j \cap G$ if and only if $\beta(u) = x_j$.

-The intervals occur in the order $a_1, a_2, \ldots, a_J$ along c.

(It may be that $a_j \cap G = \emptyset$, for some j, if there is no point $u \in c \cap G$ with $\beta(u) = x_j$.) In light of the third condition, we may furthermore isotope τ in a neighborhood of c so that $x_j \in a_j$, for each $j = 1, 2, \ldots, J$; see Figure 1.6.7a.

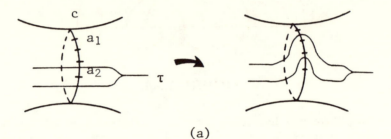

(a)

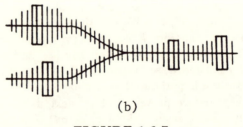

(b)

FIGURE 1.6.7

    Consider the collection of all the intervals so constructed from all the components of C. Any two such intervals are disjoint, each such interval meets G transversely, if at all (as we saw above), and each such interval intersects (transversely) exactly one branch of τ. Near each interval, we can construct a small closed disk in F foliated by parallel arcs so that each leaf in the foliation also has the properties above, and we let P denote the union of all these foliated disks. By construction, a component of $\tau - P$ is either an arc contained in some branch or contains exactly one switch of τ. One easily extends the given foliation of P to a foliated neighborhood

$(N, \mathcal{F}^\perp)$ of τ in F so that τ is the quotient of N under the collapse of leaves of $\mathcal{F}^\perp$ (as in the proof of Theorem 1.6.5); see Figure 1.6.7b.

Finally, consider a leaf g of G and a sub-arc $\alpha \subset g$ which is disjoint from P except for its endpoints: suppose that one endpoint of α lies in the leaf ℓ_+ and the other in the leaf ℓ_- of $\mathcal{F}^\perp$. By construction, $\ell_+ \neq \ell_-$, there is an imbedded smooth arc β in τ connecting ℓ_+ to ℓ_-, and α is homotopic to β keeping the endpoints on ℓ_+ and ℓ_-. Furthermore, since C hits τ efficiently, β also meets P only in its endpoints. Using that G foliates a closed subset of the compactum $F - \mathcal{U}$, it is an easy matter, as before, to further extend the foliated neighborhood $(N, \mathcal{F}^\perp)$ of τ to a larger foliated neighborhood $(N', \mathcal{F}^{\perp'})$ so that $G \subset N'$ is everywhere transverse to $\mathcal{F}^{\perp'}$. As in the proof of Theorem 1.6.5, collapsing leaves of $\mathcal{F}^{\perp'}$ produces a supporting map for the carrying $G < \tau$, and the theorem follows.   q.e.d.

§1.7 MEASURED LAMINATIONS

Suppose that L is a lamination (or geodesic lamination) in F, and let $\Lambda = \Lambda(L)$ denote the collection of all compact one-manifolds imbedded in F which are transverse to L so that the boundary (if any) of the one-manifold lies in $F - L$. A *(transverse) measure* on L is a function

$$\lambda : \Lambda \longrightarrow \left[\mathbb{R}_+ \cup \{0\} \right]$$

so that the following conditions hold.

(1) $\lambda(\alpha) = \lambda(\beta)$ whenever $\alpha \in \Lambda$ is isotopic to $\beta \in \Lambda$ through elements of Λ.

(2) λ is σ-additive in the sense that $\lambda(\alpha) = \sum_{i \in I} \lambda(\alpha_i)$ whenever $\alpha = \cup_{i \in I} \alpha_i \in \Lambda$, $\alpha_i \in \Lambda$, for all i in some countable set I, and

$$\alpha_i \cap \alpha_j = \partial \alpha_i \cap \partial \alpha_j$$

for any distinct i and j in I.

(3) The support of λ is L in the sense that for any $\alpha \in \Lambda$, $\lambda(\alpha) > 0$ if and only if $\alpha \cap L \neq \emptyset$.

If λ is a measure on the lamination L, then the pair (L, λ) is called a *measured lamination*, and if L is actually a geodesic lamination, then the pair is called a *measured geodesic lamination*. In practice, we will sometimes suppress λ from the notation and refer to L itself as a measured (geodesic) lamination.

The simplest example of a measured geodesic lamination (G, λ) is when G is a collection $\{C_j\}_{j \in J}$ of geodesic curves disjointly imbedded in F, each curve C_j is given some "weight" $\mu_j \in \mathbb{R}_+$, and the transverse measure λ of $\alpha \in \Lambda(G)$ is given by

$$\lambda(\alpha) = \sum_{j \in J} \mu_j \text{ card } (\alpha \cap C_j).$$

If (G, λ) is a measured geodesic lamination on F and $\tilde{G}$ is the full pre-image of G in $\mathbb{H}^2$, then G determines a closed, $\pi_1(F)$-invariant subset $E(\tilde{G}) \subset M_\infty$ as in Corollary 1.6.2. The transverse measure λ furthermore determines a $\pi_1(F)$-invariant measure Ξ_λ on M_∞ with support $E(\tilde{G})$ as follows. Let p be a point of $\tilde{G}$ and choose a neighborhood B of p in $\mathbb{H}^2$. There is an identification $\varphi: B \to [0, 1] \times [0, 1]$ so that $\varphi|_{\tilde{G} \cap B} = [0, 1] \times D$, where $D \subset [0, 1]$ is a closed set. Suppose that $D \cap \{0, 1\} = \emptyset$, and let $\tilde{G}_B$ be the union of all the leaves of $\tilde{G}$ which pass through B. Thus, $\tilde{G}_B$ is a closed subset of $\mathbb{H}^2$ (but is *not* $\pi_1(F)$-invariant) and, in turn, determines a closed subset $E(\tilde{G}_B) \subset M_\infty$ by Lemma 1.6.1. If $U \subset M_\infty$ is any open set so that $U \cap E(\tilde{G}) = U \cap E(\tilde{G}_B)$, then we define

$$\Xi_\lambda(U) = \lambda(\alpha),$$

where $\alpha \subset B$ is the arc $\varphi^{-1}(t \times [0, 1])$ for some $t \in [0, 1]$. It is clear that this definition is independent of all choices and defines Ξ_λ on enough open sets to determine it uniquely.

Proposition 1.7.1: *Suppose that (G, λ) is a measured geodesic lamination on F. The assignment*

$$\lambda \mapsto \Xi_\lambda$$

establishes a one-to-one correspondence between the set of all measured geodesic laminations on F and the collection of all $\pi_1(F)$-invariant measures on M_∞ whose support has the property that no member separates the elements of any other member on S^1_∞.

Proof: Given such a measure Ξ, the support of Ξ is automatically closed and corresponds to a geodesic lamination G on F by Corollary 1.6.2, and we let $\tilde{G}$ denote the full pre-image of G in $\mathbb{H}^2$. Suppose that α is an arc in F which meets G transversely and has its endpoints in $F - G$. Choose a lift $\tilde{\alpha}$ of α to $\mathbb{H}^2$, suppose that $\tilde{\alpha}$ meets the leaves $\tilde{G}_{\tilde{\alpha}}$ of $\tilde{G}$, and define

$$\lambda(\alpha) = \Xi(E(\tilde{G}_{\tilde{\alpha}})).$$

By construction, (G, λ) is a well-defined measured geodesic lamination, this assignment of (G, λ) to Ξ is a two-sided inverse to the determination of $\Xi_{\lambda'}$ from a measured geodesic lamination (G', λ') given above, and the proposition follows.   q.e.d.

Remark: The previous result gives a natural topology to the collection of all measured geodesic laminations on F; namely, as a subspace of the weak topology on measures supported on M_∞. This means that a basic open set around a measured geodesic lamination (G, λ) is described by a finite collection $f_1, \ldots, f_K \colon M_\infty \to \mathbb{R}$ of continuous functions with compact support and a positive number ε; another lamination (G', λ') lies in the corresponding open set if and only if for each $k = 1, \ldots, K$,

$$| \int_{M_\infty} f_k \, d\Xi_\lambda \; - \; \int_{M_\infty} f_k \, d\Xi_{\lambda'} | < \varepsilon.$$

Finally, notice that the measured geodesic lamination corresponding to an appropriate measure Ξ on M_∞ has compact support if and only if the support of Ξ contains no element which includes a parabolic fixed point of $\pi_1(F)$.

Requiring a geodesic lamination G to possess a transverse measure puts certain restrictions on the leaves of G, and we briefly digress to discuss this. (The sequel is independent of these considerations.)

Proposition 1.7.2: *Suppose that g is a leaf in a measured geodesic lamination of compact support. If g is a simple closed geodesic curve, then g is isolated from $G - g$. If g is a bi-infinite geodesic, then for any point $p \in G$ and any compact subarc $\alpha \subset g$ containing p, p is an accumulation point of $g - \alpha$.*

Proof: If g is a simple closed geodesic curve that is not isolated, then consider some lift $\tilde{g}$ of g to $\mathbb{H}^2$, and let $\gamma \in \pi_1(F)$ be a covering transformation which fixes $\tilde{g}$. We first show that there must be some leaf g' of G admitting a lift $\tilde{g}'$ to $\mathbb{H}^2$, so that $E(\tilde{g}) \cap E(\tilde{g}')$ is a singleton. Indeed, if not, since g is not isolated, there is a collection $\{g_i\}_{i \geq 1}$ of leaves of G and certain lifts $\tilde{g}_i$ of each g_i, so that $E(\tilde{g}_i)$ tends to $E(\tilde{g})$. Moreover, $\tilde{g}_i \cap \tilde{g} = \emptyset$ since leaves of G are disjointly imbedded, and $E(\tilde{g}_i) \cap E(\tilde{g}) = \emptyset$ by hypothesis. Since γ acts continuously on S^1_∞, we must have that $\tilde{g}_i$ intersects $\gamma(\tilde{g}_i)$ for i sufficiently large, and this contradicts that leaves of G are disjointly imbedded.

Our claim is therefore proved, and we let g' denote the leaf of G with lift $\tilde{g}'$ as above. Conjugating by an isometry of $\mathbb{H}^2$ so that $E(\tilde{g}) \cap E(\tilde{g}') = \infty \in S^1_\infty$, we observe that $\tilde{g}'$ gets progressively closer to $\tilde{g}$ as we traverse $\tilde{g}'$ towards ∞, and it follows that g' "spirals" into g as in Figure 1.7.1. Consider an arc $\alpha \in \Lambda = \Lambda(G)$ which intersects both g and g'. α contains an infinite collection $\{\alpha_i\}_{i \geq 1}$ of pairwise disjoint elements of Λ, each meeting g' once so that any two are homotopic through elements of Λ; see Figure 1.7.1. Property (1) in the definition of transverse measure gives that $\lambda(\alpha_i) = \lambda(\alpha_1)$, for all $i \geq 1$, and Property (3) gives that $\lambda(\alpha_1) \neq 0$. Finally, Property (2) gives that $\lambda(\alpha)$ must be infinite, which is absurd.

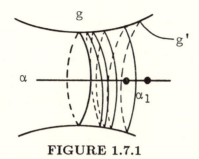

FIGURE 1.7.1

Turning to the second assertion, suppose that g is a bi-infinite geodesic and that $p \in g$ lies in a compact subarc $\alpha \subset g$ where p is isolated from $g - \alpha$. There must then be an arc $\beta \in \Lambda$ containing p so that $\beta \cap g = \{p\}$. We claim that for any point $p' \in g$ and any arc $\beta' \in \Lambda$ containing p', β' can intersect g in only finitely many points, and the proof is analogous to that of the previous paragraph. Indeed, if β' meets g in infinitely many points, then β' decomposes into infinitely many subarcs $\beta'_i \in \Lambda$, for $i \geq 1$, where $\lambda(\beta'_i) \geq \lambda(\beta) > 0$, for each i, by Properties (1) and (3) in the definition of transverse measure. Property (2) then implies that $\lambda(\beta')$ must be infinite, which is absurd, and the claim follows.

Finally, we may choose a pants decomposition of F which lies in Λ. If there is a point $p \in g$ which is isolated from $g - \alpha$, then g meets this pants decomposition in only finitely many points, as above (avoiding a neighborhood of the punctures of F, if any, by Proposition 1.6.4), and this contradicts that g is a bi-infinite geodesic.   q.e.d.

Corollary 1.7.3: *A measured geodesic lamination with compact support decomposes uniquely into a finite number of closed and isolated "minimal" sets. Each minimal set is one of the following:*

    -A closed geodesic with atomic transverse measure.

    -A measured geodesic lamination G, where each leaf of G is a bi-infinite geodesic in F which is dense in G.

Proof: This follows immediately from the previous result.   q.e.d.

Corollary 1.7.4: *If G is a measured geodesic lamination with compact support and $\alpha \in \Lambda(G)$, then $\alpha \cap G$ is the disjoint union of an isolated finite* *set and a Cantor set (i.e., closed, perfect, and no interior). Moreover, the isolated points of $\alpha \cap G$ are exactly the intersections of α with the simple closed geodesic curves in G.*

Proof: All that remains to be shown is that $\alpha \cap G$ has no interior, and this rests on the fact demonstrated in Proposition 1.6.3 above that G has zero area. Suppose that β is a closed subarc of α which lies in G. Translate β a short distance to an arc β' keeping it transverse to G with its endpoints in the same leaves of G. We claim that the interior of β' must also lie in G. If not, then choose some point $p \in \beta'$ in the complement of G, and let g^+ and g^-, respectively, be the two leaves of G closest to p along β'; since G is closed (as it is a lamination), these leaves exist. Since G consists of disjointly imbedded geodesics and the translation of β to β' is through arcs transverse to G, g^+ and g^- must intersect β in distinct points, and no point between them in β can lie in G. This contradicts that $\beta \subset G$, so $\beta \subset G$ implies that $\beta' \subset G$, and we produce a rectangle of positive area in G in contradiction to Proposition 1.6.3.   q.e.d.

Suppose that (G, λ) is a measured geodesic lamination on F with compact support and $\tau \subset F$ is a (not necessarily transversely recurrent) train track so that $G < \tau$, say with supporting map ϕ. The transverse measure λ on G determines a measure $\mu_G = \mu_{(G,\lambda)}$ on τ as follows. Suppose that b is a branch of τ, and choose a small (closed) arc α meeting τ only at a regular point of ϕ which lies in b. Thus, $\phi^{-1}(\alpha) \in \Lambda(G)$, and we define the "weight" on b to be

$$\mu_G(b) = \lambda(\phi^{-1}(\alpha)).$$

This weight $\mu_G(b)$ is independent of the choice of the arc α by Property (2) of the transverse measure on G. Assigning weights in this way to each branch of τ produces an $[\mathbb{R}_+ \cup \{0\}]$-valued function μ_G on the set of branches of τ, and another application of Property (2) shows that μ_G satisfies the switch conditions on τ, so μ_G is indeed a measure on τ.

As our notation suggests, the measure μ_G is actually independent of the supporting map ϕ as we next show.

Proposition 1.7.5: *Suppose that (G, λ) is a measured geodesic lamination with compact support carried by the train track τ. If ϕ_1 and ϕ_2 are supporting maps for the carrying $G < \tau$, then the measures μ_1 and μ_2 on τ corresponding to ϕ_1 and ϕ_2, respectively, satisfy $\mu_1 = \mu_2$.*

Proof: Let $\tilde{\phi}_i : \mathbb{H}^2 \cup S^1_\infty \to \mathbb{H}^2 \cup S^1_\infty$ denote the (extensions of the) lifts of ϕ_i to the universal cover, so $\tilde{\phi}_i$ fixes S^1_∞ pointwise, for $i = 1, 2$. If g is a leaf of G and $\tilde{g}$ a lift of g to $\mathbb{H}^2$, then consider the trainpaths $\tilde{\rho}_i = \tilde{\phi}_i(\tilde{g})$, $i = 1, 2$, on the full pre-image of τ in $\mathbb{H}^2$. Of course, both $\tilde{\rho}_1$ and $\tilde{\rho}_2$ are asymptotic to the ideal points of $\tilde{g}$. Arguing exactly as in Proposition 1.5.2, we conclude that $\tilde{\rho}_1 = \tilde{\rho}_2$.

If b is a branch of τ, then there is some point $p \in b$ which is a regular value of both ϕ_1 and ϕ_2, and we choose an arc α as above so that $\alpha \cap \tau = \{p\}$. It follows easily from the previous paragraph that $\phi_1^{-1}(\alpha)$ and $\phi_2^{-1}(\alpha)$ are isotopic through elements of $\Lambda(G)$. Thus,

$$\mu_1(b) = \lambda(\phi_1^{-1}(\alpha)) = \lambda(\phi_2^{-1}(\alpha)) = \mu_2(b),$$

where the first and third equalities reflect the definitions and the second equality follows from Property (1) in the definition of a transverse measure on a geodesic lamination.

Since b was arbitrary, we find $\mu_1 = \mu_2$, as desired.   q.e.d.

The analogue of Theorem 1.6.5 for measured geodesic laminations follows.

Corollary 1.7.6 *Every measured geodesic lamination with compact support is carried by some birecurrent train track.*

Proof: Let G be a measured geodesic lamination with compact support in F. By Theorem 1.6.5, there is some transversely recurrent train track $\tau \subset F$ so that $\tau > G$. The observation above gives a measure μ_G on τ, but it may not be that $\mu_G > 0$, so we cannot conclude that τ is recurrent. On the other hand, by Property (3) of the transverse measure on G, $\mu_G(b) > 0$ for every branch b of τ so that $\phi^{-1}(b) \cap G \neq \emptyset$. It follows that the closure of

$$\cup \{b : b \text{ is a branch of } \tau \text{ and } \mu_G(b) > 0\}$$

determines a subtrack $\tau' \subset \tau$ so that $\tau' > G$, and the measure μ_G induces a measure $\mu'_G > 0$ on τ'. By Proposition 1.3.1, τ' is recurrent, and since $\tau' \subset \tau$, τ' is transversely recurrent by Lemma 1.3.3a.   q.e.d.

Most of the remainder of this section is dedicated to providing the inverse of the construction above; namely, given a measured train track, there is a well-defined measured geodesic lamination with compact support carried by the track, and indeed, there is a unique such measured geodesic lamination so that the construction above then produces the original measured train track. We will require that the train track is transversely recurrent in what follows (and apply the material developed in §1.5).

Construction 1.7.7 Given a measured birecurrent train track (τ, μ), we shall describe the natural construction of an associated measured geodesic lamination with compact support. There are three main steps as follows: the first step is to construct a measured lamination L in F associated to (τ, μ), where $\tau > L$, and a preliminary endeavor is the construction of a certain useful bi-foliated neighborhood of τ in F; in the second step, we "straighten" L to an associated geodesic lamination $G < \tau$, and it is in this step that transverse recurrence of τ is indispensible; the third and final step is the assignment of a transverse measure to G.

Remark: Without transverse recurrence of τ, there is no guarantee of a metric on F in which L may be straightened to the geodesic lamination G. We will see in Theorem 2.7.4, however, that arbitrary measured train tracks give rise to measured geodesic laminations because any measured track may be "refined" to one which is transversely recurrent, and we may apply Construction 1.7.7 to the refined track.

We may assume that $\mu > 0$, for, if not, then

$$\tau' = \tau_\mu = \tau - \{\text{branches } b \text{ of } \tau \colon \mu(b) = 0\}$$

is a subtrack of τ (where we amalgamate each appropriate resulting bivalent switch), and the original measure μ on τ determines a positive measure $\mu' > 0$ on τ'. Since $\tau' \subset \tau$, τ' is transversely recurrent by Lemma 1.3.3a. Applying Construction 1.7.7 to the measured track (τ', μ') then associates a measured geodesic lamination to (τ, μ).

Step 1: Enumerate the branches $\{b_i\}_{i=1}^n$ of τ once and for all, and consider a collection $\{R_i\}_{i=1}^n$ of abstract (oriented) Euclidean rectangles with the same index set; the rectangle R_i is taken to have length one and width $\mu(b_i)$, and we choose, once and for all, a vertical segment ξ_i in the middle of R_i, for each i. Each R_i is endowed with two canonical foliations $\mathcal{F}_i$ (and $\mathcal{F}_i^\perp$, respectively) with leaves perpendicular (parallel) to ξ_i, and the boundary ∂R_i decomposes into two "horizontal" arcs (which are perpendicular to ξ_i and of length one) and two "vertical" arcs (which are parallel to ξ_i and of length $\mu(b_i)$); see Figure 1.7.2. Each branch b_i of τ has exactly two ends, and we may index the vertical arcs in ∂R_i by the ends of b_i; each such vertical arc inherits an orientation from that of the rectangle R_i.

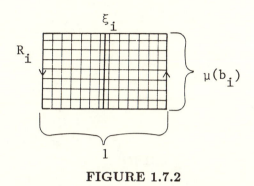

FIGURE 1.7.2

We define a bi-foliated surface which is naturally homeomorphic to a closed regular neighborhood N of τ in F by identifying certain sets of vertical arcs in the boundaries of rectangles as follows. Suppose that $v \in F$ is a switch of τ with $e_1, \ldots, e_r$ the incoming ends of branches incident on v and $e_{r+1}, \ldots, e_{r+t}$ outgoing. We may suppose that the indices are chosen so that as one traverses a small simple closed curve in F encircling v (in the clockwise direction), one encounters the half-branches of τ corresponding to the ends $e_1, \ldots, e_r, e_{r+1}, \ldots, e_{r+t}$ in this order; let $a_1, \ldots, a_r, a_{r+1}, \ldots, a_{r+t}$ denote the corresponding oriented vertical arcs. By the switch conditions, the sum ζ of the Euclidean lengths of $\{a_i\}_{i=1}^r$ agrees with the sum of the Euclidean lengths of $\{a_j\}_{j=r+1}^{r+t}$, and there are unique Euclidean-length-preserving maps

$$f: \coprod_{i=1}^{r} a_i \to [0, \zeta]$$

and

$$g: \coprod_{j=r+1}^{r+t} a_j \to [0, \zeta],$$

where f is orientation-reversing and maps the initial point of a_1 to ζ, g is orientation-preserving and maps the endpoint of a_{r+1} to ζ,

$$f(\partial a_i) \cap f(\partial a_{i+1}) \neq \emptyset, \text{ for } i = 1, \ldots, r-1,$$

and

$$g(\partial a_j) \cap g(\partial a_{j+1}) \neq \emptyset, \text{ for } j = r+1, \ldots, r+t-1.$$

Finally, each fiber of the map $f \coprod g$ is collapsed to a (distinct) point; see Figure 1.7.3.

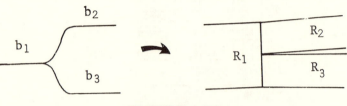

FIGURE 1.7.3

Take the quotient of $\coprod_{i=1}^{n} R_i$ by these identifications as v ranges over all the switches of τ. The result is a surface which we henceforth identify with a neighborhood N of τ in F in order to simplify notation. The foliations $\{\mathcal{F}_i\}_{i=1}^{n}$ and $\{\mathcal{F}_i^{\perp}\}_{i=1}^{n}$ combine in the natural way to produce the transverse foliations $\mathcal{F}$ and $\mathcal{F}^{\perp}$, respectively, and each of $\mathcal{F}$ and $\mathcal{F}^{\perp}$ inherits a transverse measure (in the sense of [FLP]) from the Euclidean metrics on the rectangles.

A point $p \in \partial N$ is called a *singular point* if the fiber of the natural projection $\coprod_{i=1}^{n} R_i \to N$ over p has cardinality greater than two; the singular points are thus the "cusps" of ∂N. A leaf of $\mathcal{F}^{\perp}$ is a *tie*, and it is a *singular tie* if it moreover contains a singular point. If Σ is the union of all the leaves of $\mathcal{F}$ which contain singular points and Π is the set of all singular points, then each "component" of $\Sigma - \Pi$ is called a *singular leaf* of $\mathcal{F}$, and any leaf of $\mathcal{F}$ other than a singular leaf is a *non-singular leaf*. If ℓ is a leaf of $\mathcal{F}$ which is not a simple closed curve and $p \in \ell$, then each component of $\ell - p$ is called a *half-leaf* of ℓ. Thus, any half-leaf of a non-singular leaf has infinite $\mathcal{F}^{\perp}$-measure. In contrast, a singular leaf is called *finite* if every half-leaf of it has finite $\mathcal{F}^{\perp}$-measure and is called *semi-infinite* otherwise. Notice that if ℓ is a semi-infinite singular leaf and $p \in \ell$, then exactly one component of $\ell - p$ has finite $\mathcal{F}^{\perp}$-measure. Finally, notice that each singular leaf must lie either entirely in the interior of N or entirely in ∂N; furthermore, a semi-infinite singular leaf is always of the former type while a finite singular leaf might be of either type.

This completes the required construction of our bi-foliated neighborhood $(N, \mathcal{F}, \mathcal{F}^{\perp})$ of τ in F. (This neighborhood is technically useful both in the rest of this section and elsewhere in this work.) The idea of the construction of a lamination L associated to $(N, \mathcal{F}, \mathcal{F}^{\perp})$ is now easy to describe: alter the foliation $\mathcal{F}$ by "splitting" along all of the singular leaves, so that only "packets" of parallel non-singular leaves remain. These leaves form the lamination L.

To make this precise, let θ_i denote the intersection of ξ_i with the collection of all the singular leaves of $\mathcal{F}$, and notice that θ_i is either a finite or a countably infinite set (and $\theta_i = \emptyset$ if and only if b_i corresponds to a curve component of τ). For each $i = 1, \ldots, n$, we will construct an interval $\eta_i \subset \mathbb{R}$ together with a closed subset $K_i \subset \eta_i$ and a continuous map $\iota_i : \eta_i \to \xi_i$ so that the following conditions hold.

(i) Letting m denote the Lebesgue measure on $\mathbb{R}$, we have $m(K_i) = \mu(b_i) = m(\xi_i)$, and $m(\eta_i) \leq 2m(\xi_i)$.

(ii) The restriction $\iota_i \mid_{K_i}$ is one-to-one over points of $\partial\xi_i \cup (\xi_i - \theta_i)$.

(iii) The restriction $\iota_i \mid_{K_i}$ is two-to-one over points of $\theta_i - \partial\xi_i$.

To construct such intervals and maps, enumerate the set $\theta_i \subset \xi_i$ as $\{x_{ij}\}_{j=1}^{n_i}$ (where perhaps $n_i = \infty$), and let m_i denote the measure on ξ_i given by adding to the Lebesgue measure m each Dirac measure with mass $2^{-j}m(\xi_i)$ at x_{ij}, for each $j = 1, \ldots, n_i$. For each $i = 1, \ldots, n$ let η_i be the interval of length

$$m(\eta_i) = m(\xi_i)\Big\{1 + \sum_{j=1}^{n_i} 2^{-j}\Big\} \leq 2m(\xi_i).$$

The distribution of m_i maps ξ_i to η_i and is discontinuous exactly on θ_i. The inverse distribution $\iota_i : \eta_i \to \xi_i$, however, is continuous. Let $K_i \subset \eta_i$ be the complement of the interior of the intervals (which map to θ_i) on which ι_i is constant. It is evident that the three demands above are met.

Fix a switch v, and recall the intervals and maps

$$f : \coprod_{j=1}^{r} a_j \to [0, \zeta] \text{ and } g : \coprod_{j=r+1}^{r+t} a_j \to [0, \zeta],$$

which arose in the construction of the neighborhood N. Now consider a new collection of abstract Euclidean rectangles $\{R_i'\}_{i=1}^{n}$ where R_i' has length one and width $2\mu(b_i)$. Let $a_1', \ldots, a_r', a_{r+1}', \ldots, a_{r+t}'$ denote the corresponding vertical arcs in $\{\partial R_i'\}$, and

$$f' : \coprod_{j=1}^{r} a_j' \to [0, 2\zeta] \text{ and } g' : \coprod_{j=r+1}^{r+t} a_j' \to [0, 2\zeta],$$

the corresponding maps. Using the conditions above, one sees that there is a unique measure-preserving homeomorphism

$$\Phi : \coprod_{j=1}^{r} K_j \to \coprod_{j=r+1}^{r+t} K_j$$

so that

$$f \circ \Pi \circ \Big(\coprod_{j=1}^{r} \iota_j\Big) = g \circ \Pi \circ \Big(\coprod_{j=r+1}^{r+t} \iota_j\Big) \circ \Phi,$$

where Π denotes the disjoint union of the natural projections $\xi_j \to a_j$ for a_j a vertical arc in ∂R_i. Since $m(\eta_i) < 2\mu(b_i)$ by Condition (i) above, we may regard the rectangle $\eta_i \times I$ as horizontally imbedded in R'_i, where I denotes the unit interval; via this identification, we may regard Φ as a map

$$\Phi \colon f' \circ \Pi'\Big(\coprod_{j=1}^{r} K_j\Big) \subset [0, 2\zeta] \to g' \circ \Pi'\Big(\coprod_{j=r+1}^{r+t} K_j\Big) \subset [0, 2\zeta],$$

where Π' denotes the disjoint union of the natural projections, as before. Φ extends easily to a homeomorphism $\hat{\Phi} \colon [0, 2\zeta] \to [0, 2\zeta]$.

Finally, construct a neighborhood $\hat{N}$ of τ in F using the map $g' \circ \hat{\Phi}$ in place of g' in the identification of vertical arcs near v, for each switch v of τ. By construction, the collection of arcs $\{K_i \times I\}_{i=1}^{n}$ combine to give a foliation L of some subset of $\hat{N}$, and this subset is closed (so L is actually a lamination) since cross-sections of L are closed by construction. The vertical foliations of the rectangles R'_i combine as before to produce a tie neighborhood of τ which is everywhere transverse to L, so in fact $L < \tau$.

There is a canonical transverse measure λ on L defined as follows. Let Λ denote the collection of arcs transverse to L as in the definition of transverse measure, and let $\pi_i \colon \eta_i \times I \to \eta_i$ denote the horizontal projection. Define

$$\lambda \colon \Lambda \to [\mathbb{R}_+ \cup \{0\}]$$
$$: \alpha \mapsto m(K_i \cap \pi_i(\alpha \cap (\eta_i \times I))) = m(\iota_i \circ \pi_i(\alpha \cap (\eta_i \times I))),$$

so that Properties (1)-(3) in the definition of transverse measure evidently hold.

This finishes the construction of the measured lamination (L, λ), and Step 1 is complete. We add parenthetically that since $L < \tau$, there is an induced measure $\mu_{(L,\lambda)}$ on τ, and $\mu_{(L,\lambda)}$ is the original measure μ by construction.

Step 2: Next, we replace L by a geodesic lamination. The idea is that *because τ is transversely recurrent*, the results of §1.5 apply to show that each lift $\tilde{\ell}$ of a leaf ℓ of L to the universal cover $\mathbb{H}^2$ of F has two distinct well-defined limit points in S^1_∞; let $\tilde{\ell}'$ denote the geodesic in $\mathbb{H}^2$ with these limit points as its ideal points. The collection $\{\tilde{\ell}' \colon \ell \in L\}$ will descend to a geodesic lamination on F, as we shall see.

More precisely, each leaf ℓ of L determines a bi-infinite trainpath $\tilde{\rho}$ on the full pre-image $\tilde{\tau}$ of τ in $\mathbb{H}^2$. (In case ℓ is a simple closed curve, imagine traversing ℓ repeatedly to define a bi-infinite trainpath.) The trainpath $\tilde{\rho}$ determines a point $E(\tilde{\rho}) \in M_\infty$ by Proposition 1.5.1, and there is a unique geodesic $\tilde{\ell}'$ in $\mathbb{H}^2$ so that $E(\tilde{\rho}) = E(\tilde{\ell}')$. $\tilde{\ell}'$ projects, in turn, to a geodesic ℓ' in F. Clearly, if $\tilde{\ell}_1$ and $\tilde{\ell}_2$ are each lifts of a common leaf ℓ of L, then $\tilde{\ell}_1'$ and $\tilde{\ell}_2'$ project to the same geodesic ℓ' in F, so it makes sense to replace each leaf ℓ of L by the geodesic ℓ' described above. In so doing, we produce a collection G of geodesics in F, and the full pre-image of G in $\mathbb{H}^2$ is denoted $\tilde{G}$. Notice that many leaves of L amalgamate to form one geodesic in G.

Theorem 1.7.8: *The collection G of geodesics given by Construction 1.7.7 forms a geodesic lamination with compact support, and $G < \tau$.*

Proof: If $\tilde{\rho}_1$ and $\tilde{\rho}_2$ are each bi-infinite trainpaths on $\tilde{\tau}$ corresponding to leaves of L, then $E(\tilde{\rho}_1)$ cannot separate the members of $E(\tilde{\rho}_2)$ since the leaves of L are disjointly imbedded. The same therefore follows for any $E(\tilde{g}_1)$ and $E(\tilde{g}_2)$, where $\tilde{g}_1$ and $\tilde{g}_2$ are geodesics in $\tilde{G}$. It follows that G is a collection of geodesics disjointly imbedded in F. The difficulty lies in showing that G foliates a *closed* subset of F, or equivalently (by Lemma 1.6.1), that $E(\tilde{G})$ is a closed subset of M_∞. We will actually describe $E(\tilde{G})$ as a nested intersection of closed subsets of M_∞, each closed subset arising from some train track.

Recall from Theorem 1.5.4 that

$$E(\tau) = \{E(\tilde{\rho}) : \rho \text{ is a bi} - \text{infinite trainpath on } \tilde{\tau}\}$$

is a closed subset of M_∞. Of course, $E(\tilde{G})$ is a subset of $E(\tau)$, but it is much smaller. We will identify

$$E(\tilde{G}) = \bigcap_{m \geq 0} E(\tau_m)$$

as a nested intersection, where each τ_m is a transversely recurrent train track carrying L with $\tau_0 > \tau_1 > \ldots$, to conclude that $E(\tilde{G}) \subset M_\infty$ is indeed closed.

To this end, recall the bi-foliated neighborhood $(N, \mathcal{F}, \mathcal{F}^\perp)$ of Step 1, and let $\{\varsigma_k\}_{k=1}^K$ denote the singular leaves of $\mathcal{F}$ which lie in the interior of N. Suppose that each ς_k, for $k = 1, \ldots, K'$, is a semi-infinite singular leaf issuing forth from a singular point s_k, and each ς_k, for $k = K' + 1, \ldots, K$, is a finite singular leaf (where $K' = K$ if there are no finite and $K' = 0$ if there are no semi-infinite singular leaves). For each $k = 1, \ldots, K'$, the collection of all singular ties of $(N, \mathcal{F}, \mathcal{F}^\perp)$ decompose ς_k into a semi-infinite collection of open intervals, which we linearly order from the order in which they are traversed by ς_k; ς_k is given its natural orientation starting from s_k. For each $m \geq 1$ and each $k = 1, \ldots K'$, let $\varsigma_k(m)$ denote the smallest connected subset of $\varsigma_k \cup \{s_k\}$ containing both s_k and the mth open interval in this decomposition (as in Figure 1.7.4a); thus, each $\varsigma_k(m)$ is a half-open finite arc with one endpoint (namely s_k) in ∂N.

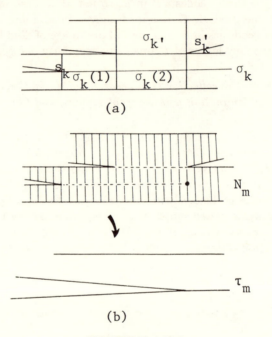

(a)

(b)

FIGURE 1.7.4

Consider the subset

$$N_m = N - \bigcup_{k=K'+1}^{K} \bar{\varsigma}_k - \bigcup_{k=1}^{K'} \varsigma_k(m)$$

of N, where $m \geq 1$ and $\bar{\varsigma}$ denotes the closure in F of a finite path ς. Notice that the foliations $\mathcal{F}$ and $\mathcal{F}^{\perp}$ restrict to foliations $\mathcal{F}_m$ and $\mathcal{F}_m^{\perp}$ of N_m, respectively. A leaf of $\mathcal{F}_m^{\perp}$ which contains a point of $\bar{\varsigma}_k(m) - \varsigma_k(m)$ is called a "singular tie" of $\mathcal{F}_m^{\perp}$. As in the proof of Theorem 1.6.5, collapsing each leaf of $\mathcal{F}_m^{\perp}$ to a point produces a train track $\tau_m \subset N_m$, where the singular ties of $\mathcal{F}_m^{\perp}$ give rise to switches of τ_m: see Figure 1.74b. Using this collapse, one easily shows that $\tau_m > \tau_{m+1}$. At the same time, the collapse of ties of $\mathcal{F}^{\perp}$ itself produces the original track τ, and it follows similarly that $\tau = \tau_0 > \tau_m$, for each $m \geq 1$.

Since τ_0 is transversely recurrent by hypothesis and, moreover, $\tau_0 > \tau_m$, Lemma 1.3.3 part (b) applies to show that τ_m is also transversely recurrent, for each m. It follows that each $E(\tau_m) \subset \mathrm{M}_\infty$ is closed by Theorem 1.5.4, and since $\tau_m > \tau_{m+1}$, we find that $E(\tau_m) \supset E(\tau_{m+1})$, for $m \geq 0$, by Proposition 1.5.5. The intersection

$$\mathcal{E} = \bigcap_{m \geq 0} E(\tau_m) \subset \mathrm{M}_\infty$$

is therefore nested, and $\mathcal{E}$ is a closed subset of M_∞. Each τ_m carries L by construction, so $E(\tau_m) \supset E(\tilde{G})$ by definition, and it follows that $E(\tilde{G}) \subset \mathcal{E}$.

To prove the reverse inclusion requires some work. It is convenient to introduce the following notion: suppose that $\alpha: [0, I] \to N$ for $I \in \mathbb{Z}_+$, is an arc in N which is transverse to the ties of $\mathcal{F}^\perp$ so that each point $\alpha(i)$, for $i \in [0, I] \cap \mathbb{Z}$, lies in a singular tie of $\mathcal{F}^\perp$, and these are the only intersections of α with singular ties of $\mathcal{F}^\perp$. Such a path α is called a *(finite) tie-transverse path* in N of length I, and two tie-transverse paths are regarded as equivalent if they are *homotopic along $\mathcal{F}^\perp$* in the sense that there is a homotopy from one to the other so that the track of any point (under the isotopy) is contained in some tie of N. There is a natural one-to-one correspondence between the set of all (equivalence classes of) tie-transverse paths of length I in N and the set of all (equivalence classes of) finite trainpaths on τ of length I; this correspondence is induced by the correspondence between the branches of τ and the component rectangles of N.

The following result gives the crucial estimate in the proof below that $\mathcal{E} \subset E(\tilde{G})$ and is of interest in its own right.

Lemma 1.7.9: *Suppose that ρ is a trainpath on τ of length at most $2m+1$ and the tie-transverse path of ρ is homotopic along the leaves of $\mathcal{F}^\perp$ to a path contained in N_m, for some $m \geq 0$. Then ρ is realized by a leaf ℓ of L in the sense that ρ arises by restricting the domain of a trainpath on τ corresponding to ℓ.*

Proof: By definition of the lamination L, it suffices to prove that ρ is realized by some non-singular leaf ℓ of $\mathcal{F}$. Let the length of ρ be j, where $1 \leq j \leq 2m + 1$, and proceed by induction on j. For the basis step $j = 1$, suppose that b is a branch of τ and let ξ be a tie of $\mathcal{F}^\perp$ meeting b. Since the intersection θ of ξ with the collection of all singular leaves of $\mathcal{F}$ is countable, we must have $\xi - \theta \neq \emptyset$. Simply choose any leaf ℓ of $\mathcal{F}$ meeting $\xi - \theta$ to complete the basis step.

For the inductive step, pass to the universal cover $\mathbb{H}^2$ of F and consider the full pre-image $(\tilde{N}, \tilde{\mathcal{F}}, \tilde{\mathcal{F}}^\perp)$ in $\mathbb{H}^2$ of the bi-foliated neighborhood

$(N, \mathcal{F}, \mathcal{F}^\perp)$. Let $\tilde{\rho} \colon [0, j] \to \tilde{N}$ be a lift to $\mathbb{H}^2$ of a trainpath ρ on τ of length j, where $2m + 1 \geq j \geq 2$, so that $\rho \mid_{[0, j-1]}$ and $\rho \mid_{[j-1, j]}$, respectively, are realized by leaves ℓ_1 and ℓ_2 of $\mathcal{F}$ by the inductive hypothesis. Let $\tilde{\ell}_1' \subset \tilde{\ell}_1$ and $\tilde{\ell}_2' \subset \tilde{\ell}_2$, respectively, denote lifts of the corresponding finite tie-transverse paths so that the terminal point $\tilde{p}$ of $\tilde{\ell}_1'$ and the initial point $\tilde{q}$ of $\tilde{\ell}_2'$ lie on the same singular tie $\tilde{t}_0$ of $\tilde{\mathcal{F}}^\perp$. Suppose, for definiteness, that $\tilde{\ell}_1'$ lies to the left of $\tilde{p}$ and $\tilde{\ell}_2'$ lies to the right of $\tilde{q}$ near $\tilde{t}_0$. There may be singular points of $\tilde{\mathcal{F}}$ on $\tilde{t}_0$ between $\tilde{p}$ and $\tilde{q}$, and we distinguish two cases: such a singular point $\tilde{s}_0$ is called a "right cusp" if the singular leaf of $\tilde{\mathcal{F}}$ issuing from $\tilde{s}_0$ lies to the right of $\tilde{t}_0$ near $\tilde{s}_0$; $\tilde{s}_0$ is called a "left cusp" in the contrary case. (See Figure 1.7.5a.)

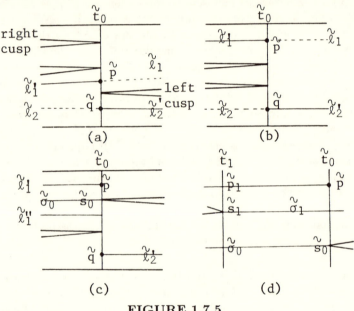

FIGURE 1.7.5

We finish the proof of the lemma by induction on the number c of left cusps on $\tilde{t}_0$ between $\tilde{p}$ and $\tilde{q}$. If $c = 0$, then ℓ_1 actually realizes ρ; see Figure 1.7.5b. It follows that a collection of nearby leaves also realize ρ, and the proof is completed by a cardinality argument as before. If $c \geq 1$, let $\tilde{s}_0 \in \tilde{t}_0$ denote the left cusp nearest $\tilde{p}$ lying between $\tilde{p}$ and $\tilde{q}$ on $\tilde{t}_0$, and let $\tilde{\sigma}_0$ denote the singular leaf of $\tilde{\mathcal{F}}$ issuing forth from $\tilde{s}_0$. We claim that $\tilde{\ell}_1'$ is homotopic along ties of $\tilde{\mathcal{F}}^\perp$ to a tie-transverse path $\tilde{\ell}_1''$, which is contained in some leaf of $\tilde{\mathcal{F}}$, so that the endpoint of $\tilde{\ell}_1''$ on $\tilde{t}_0$ corresponding to $\tilde{p}$ lies between $\tilde{s}_0$ and $\tilde{q}$; see Figure 1.75c. Continuing in this way, we reduce to the case $c = 0$, and the proof is completed.

To prove our claim, first note that there can be no right cusp on $\tilde{t}_0$

between $\tilde{p}$ and $\tilde{s}_0$; for otherwise, there could be no tie-transverse path in $\tilde{N}$ which corresponds to the original trainpath ρ, and this is contrary to hypothesis. Thus, the arc in $\tilde{t}_0$ between $\tilde{p}$ and $\tilde{s}_0$ is free from singular points of $\tilde{\mathcal{F}}$.

Let $\tilde{p}_1$ denote the point of $\tilde{\ell}_1' - \{\tilde{p}\}$ nearest $\tilde{p}$ along $\tilde{\ell}_1'$ so that the tie $\tilde{t}_1$ of $\tilde{\mathcal{F}}^{\perp}$ through $\tilde{p}_1$ contains a singular point $\tilde{s}_1$ of $\tilde{\mathcal{F}}$ which lies between $\tilde{p}_1$ and $\tilde{\sigma}_0 \cap \tilde{t}_1$ on $\tilde{t}_1$; see Figure 1.7.5d. If there is no such point $\tilde{p}_1$, then $\tilde{\ell}_1'$ is homotopic along ties of $\tilde{\mathcal{F}}^{\perp}$ to a subset of $\tilde{\sigma}_0$, and one easily produces the required tie-transverse path $\tilde{\ell}_1''$. As before, the singular leaf issuing forth from $\tilde{s}_1$ cannot lie to the left of $\tilde{t}_1$ near $\tilde{s}_1$ since otherwise there could be no tie-transverse path in N corresponding to ρ. Thus, the singular leaf $\tilde{\sigma}_1$ of $\tilde{\mathcal{F}}$ issuing forth from $\tilde{s}_1$ must lie to the right of $\tilde{t}_1$ near $\tilde{s}_1$.

It cannot be that $\tilde{\sigma}_0 = \tilde{\sigma}_1$ since this would violate that ρ is homotopic along ties of $\mathcal{F}^{\perp}$ to a path in $N_m \subset N$; in the same way, one may conclude that $\tilde{\sigma}_0$ and $\tilde{\sigma}_1$ are each semi-infinite leaves of $\tilde{\mathcal{F}}$. Since the length of $\tilde{\ell}_1'$ is at most $2m$, as we traverse $\tilde{\ell}_1'$ starting from $\tilde{p}$, we first encounter the tie containing the terminal point of $\tilde{\sigma}_1(m)$ and then encounter the tie containing the terminal point of $\tilde{\sigma}_0(m)$ (or, perhaps, these terminal points lie on a common tie). In any case, this again contradicts our hypothesis on ρ. We have therefore established that there is no such point $\tilde{p}_1 \in \tilde{\ell}_1'$, and the proof is complete. q.e.d.

Returning now to the proof of Theorem 1.7.8, we must show that $\mathcal{E} \subset E(\tilde{G})$. Suppose that $\{x, y\} \in \mathcal{E} \subset M_\infty$, and let $\tilde{\rho}_0$ be a trainpath on $\tilde{\tau}_0$ so that $E(\tilde{\rho}_0) = \{x, y\}$ with projection ρ_0 to F. Suppose that $\rho_0((0, 1))$ is the branch b of τ, and recall the tie ξ of N corresponding to b, the map $\iota : \eta \to \xi$, and the closed set $K \subset \eta$ described in Step 1. If $z \in K$, then consider the leaf ℓ_z of L through $\iota(z)$; ℓ_z determines a corresponding bi-infinite trainpath ρ_z on τ. For each $k \geq 1$, define

$$J_k = \{z \in K : \rho_z \mid_{[-k, k]} = \rho_0 \mid_{[-k, k]}\}$$

so that each $J_k \subset K \subset \eta$ is closed and $J_1 \supset J_2 \supset \dots$ by construction.

We must show that $J_k \neq \emptyset$ for $k \geq 1$, so that by compactness of η, there is some $z_0 \in \cap_{k \geq 1} J_k$. For in this case, the trainpath ρ_{z_0} on τ lifts to the trainpath $\tilde{\rho}_0$ on $\tilde{\tau}$, and since ρ_{z_0} arises from the leaf ℓ_{z_0} of L, it follows that $E(\tilde{\rho}_0) = \{x, y\} \in E(\tilde{G})$, as desired.

To prove this final claim, since $\{x, y\} \in \mathcal{E}$, there is some bi-infinite trainpath $\tilde{\rho}_m$ on $\tilde{\tau}_m$ with $E(\tilde{\rho}_m) = \{x, y\}$ for each $m \geq 1$. Insofar as $\tau_m < \tau_0$ for $m \geq 0$ (as we saw above), Proposition 1.5.5 applies, so $\tilde{\rho}_m$ gives rise to a trainpath $\tilde{\rho}_m'$ on $\tilde{\tau}_0$ where $E(\tilde{\rho}_m') = E(\tilde{\rho}_m) = \{x, y\}$. By Proposition 1.5.2, there must be an equality $\tilde{\rho}_0 = \tilde{\rho}_m'$ of trainpaths (up to reversal) for $m \geq 0$. It follows that for each m, the restriction $\rho_0 \mid_{[-m, m+1]}$

satisfies the hypotheses of Lemma 1.7.9, and the corresponding leaf gives rise to a point of J_m. Thus, $J_m \neq \emptyset$ for all $m \geq 1$, and this completes the proof that $\mathcal{E} \subset E(\tilde{G})$.

It follows that $\mathcal{E} = E(\tilde{G}) \subset M_\infty$ is closed, so G is a geodesic lamination. Finally, since $E(\tilde{G}) = \mathcal{E} \subset E(\tau_m)$, Theorem 1.6.6 implies that $G < \tau_m$, for all $m \geq 0$. In particular, $G < \tau_0 = \tau$, and G has compact support by construction, so the proof of Theorem 1.7.8 is completed.   q.e.d.

We record here the following immediate corollary of the argument above.

Corollary 1.7.10: *Suppose that (τ, μ) is a birecurrent train track giving rise to the geodesic lamination $G \subset F$ and $\alpha \subset N$ is a bi-infinite tie-transverse path which is disjoint from the singular leaves of $\mathcal{F}$. If $\tilde{\rho}$ denotes a trainpath on the full pre-image $\tilde{\tau} \subset \mathbb{H}^2$ corresponding to α, then $E(\tilde{\rho}) \in E(\tilde{G})$.*

This finishes our discussion of the geodesic lamination G, and Step 2 is complete.

Step 3: It remains to define a transverse measure λ' on G, and this is easily done. Let Λ' denote the collection of arcs transverse to G with endpoints in $F - G$. We claim that any arc $\alpha' \in \Lambda'$ can be written as the concatenation

$$\alpha' = \alpha'_1 * \alpha'_2 * \ldots * \alpha'_K,$$

where each arc α'_k meets the lamination G efficiently in the sense that there is no bigon imbedded in F whose frontier consists of one C^1 segment contained in α'_k and one contained in a leaf of G. To prove the claim, we suppose the contrary, so there are infinitely many disjoint sub-arcs of α which do not meet G efficiently; these sub-arcs have an accumulation point, say $x \in \alpha'$, and x is necessarily a point of $\alpha' \cap G$ since each sub-arc contains a point of $\alpha' \cap G$, and G is closed. This violates transversality of α' with G at the point x and proves our claim.

Thus, by Property (2) in the definition of transverse measure, it suffices to define λ' on the arcs in Λ' which meet G efficiently. Recall the lamination L constructed in Step 1 together with its transverse measure λ defined on the family Λ of arcs. An efficient arc $\alpha' \in \Lambda'$ gives rise in the natural way to an arc $\alpha \in \Lambda$ which meets L efficiently, and we define

$$\lambda'(\alpha') = \lambda(\alpha).$$

The required properties of λ' follow directly from the corresponding properties of λ.

This finishes the construction of a transverse measure Λ' on G, so Step (3) is complete. Thus, we have described the construction of a measured geodesic lamination (G, λ') from a measured recurrent train track (τ, μ). To finish our discussion, we prove that Construction 1.7.7 is actually the two-sided inverse of the construction of the measure $\mu_{(G,\lambda)}$ on τ from (G, λ) when $G < \tau$, thus canonically identifying the collection of all measured geodesic laminations of compact support carried by τ with the collection of all measures on τ.

Lemma 1.7.11: *Suppose that $\tau \subset F$ is a transversely recurrent train track.*

> *(a) If μ_1 and μ_2 are measures on τ giving rise to the same measured geodesic lamination (G, λ), then $\mu_1 = \mu_2$.*

> *(b) If (G_i, λ_i) are measured geodesic laminations on F, with $G_i < \tau$ giving rise to the same measure $\mu = \mu_{(G_i,\lambda_i)}$, for $i = 1, 2$, then $(G_1, \lambda_1) = (G_2, \lambda_2)$.*

Proof: For part (a), choose a supporting map ϕ for the carrying $G < \tau$, fix a branch b of τ, and choose a point $p \in b$ which is a regular value of ϕ. Choose an arc α_b crossing b transversely at p and meeting τ nowhere else, and let $\alpha'_b = \phi^{-1}(\alpha_b)$. By construction and Proposition 1.7.5, $\mu_1(b) = \lambda(\alpha'_b) = \mu_2(b)$, as desired.

For part (b), recall the construction of the sequence $\tau = \tau_0 > \tau_1 > \dots$ of birecurrent train tracks built from the measure μ on τ as in the proof of Theorem 1.7.8. As shown there

$$E(\tilde{G}_1) = \cap_{m \geq 0} E(\tau_m) = E(\tilde{G}_2),$$

where $\tilde{G}_i$ denotes the full pre-image of G_i in $\mathbb{H}^2$. We conclude that $G_1 = G_2$, as was claimed.   q.e.d.

The *measured lamination space* $\mathcal{ML}(F)$ of a surface F is the collection of all measured geodesic laminations in F, where F has some fixed hyperbolic structure. The topology on $\mathcal{ML}(F)$ is induced from the weak topology on measures supported on M_∞ as in the remark following Proposition 1.7.1. This topology (and indeed $\mathcal{ML}(F)$ itself) seems to depend on the hyperbolic structure on the surface F. However, Nielsen's Extension Theorem says that the lift to $\mathbb{H}^2$ of a continuous mapping $F_1 \to F_2$ between two hyperbolic surfaces extends continuously to S^1_∞; thus, a homeomorphism of F induces a homeomorphism $M_\infty \to M_\infty$, which may be used to naturally identify $\mathcal{ML}(F_1)$ with $\mathcal{ML}(F_2)$. Hence it is legitimate to write simply $\mathcal{ML}(F)$ for the collection of all measured geodesic laminations in F without any reference to the underlying hyperbolic structure on F.

Another description of the topology of $\mathcal{ML}(F)$ (which is often useful) is as follows. Suppose that $(G, \lambda) \in \mathcal{ML}(F)$, let $\varepsilon \in \mathbb{R}_+$, and let $\{\alpha_k\}_1^K$ be a finite collection of arcs lying in $\Lambda(G)$. A basic open set in $\mathcal{ML}(F)$ about (G, λ) is given by the collection of all $(G', \lambda') \in \mathcal{ML}(F)$ so that

$$\alpha_k \in \Lambda(G') \text{ and } |\lambda(\alpha_k) - \lambda'(\alpha_k)| < \varepsilon$$

for all $k = 1, \ldots, K$. Using the assignment $(G, \lambda) \mapsto \Xi_\lambda$ defined before Proposition 1.7.1, one sees without difficulty that that the two topologies defined on $\mathcal{ML}(F)$ actually coincide.

We define $\mathcal{ML}_0(F) \subset \mathcal{ML}(F)$ to be the subspace of measured geodesic laminations in F with compact support. Elements of $\mathcal{ML}_0(F)$ lie outside the unit horoball neighborhood of the punctures of F by Proposition 1.6.4.

Multiplying the transverse measure on a geodesic lamination by a positive constant gives an $\mathbb{R}_+$-action on each of $\mathcal{ML}(F)$ and $\mathcal{ML}_0(F)$, and this action is evidently properly discontinuous. The *space of projective laminations* in F is defined to be $\mathcal{PL}(F) = (\mathcal{ML}(F) - \{0\})/\mathbb{R}_+$, where 0 denotes the empty lamination, and the *space of projective laminations with compact support* in F is similarly defined by $\mathcal{PL}_0(F) = (\mathcal{ML}_0(F) - \{0\})/\mathbb{R}_+$. The topologies on $\mathcal{PL}(F)$ and $\mathcal{PL}_0(F)$ are the quotient topologies inherited from those on $\mathcal{ML}(F)$ and $\mathcal{ML}_0(F)$ respectively.

If $\tau \subset F$ is a birecurrent train track, then the collection $V(\tau)$ of all transverse measures has a natural topology, where two measures are close if the weights they assign to each branch are close. (The structure of $V(\tau)$ will be further investigated in §2.1 and §2.2.) Thus, we may regard Construction 1.7.7 as a map from the topological space $V(\tau)$ to the topological space $\mathcal{ML}_0(F)$. Collecting our results about Construction 1.7.7 into a theorem for later use, we have

Theorem 1.7.12: *Fix a birecurrent train track $\tau \subset F$ and let $V(\tau)$ be the space of transverse measures on τ. Construction 1.7.7 establishes a continuous injection $V(\tau) \to \mathcal{ML}_0(F)$ whose image consists of all the measured geodesic laminations of compact support carried by τ.*

Remark: We will find (in Theorem 2.7.4) that the hypothesis of transverse recurrence on τ can be dropped.

Proof: By Lemma 1.7.11, only continuity of the map induced by Construction 1.7.7 requires comment, and we use the description of the topology on $\mathcal{ML}_0(F)$ in terms of arcs in the surface given above. Suppose that (τ, μ) gives rise to the the measured geodesic lamination (G, λ') and let (L, λ)

be the corresponding measured lamination defined in Step 1 of Construction 1.7.7. Given an arc $\alpha' \in \Lambda(G)$, we may assume that α' meets G efficiently as in Step 3 of Construction 1.7.7, so α' gives rise to an arc $\alpha \in \Lambda(L)$. Finally, $\lambda'(\alpha')$ is defined to agree with $\lambda(\alpha)$, and this latter quantity is obviously a continuous function of μ. Thus, the map induced by Construction 1.7.7 is continuous, as was asserted.   q.e.d.

Corollary 1.7.13: *Suppose that $G \subset F$ is a measured geodesic lamination carried by the birecurrent train track $\tau \subset F$. There is a sequence $\tau = \tau_0 > \tau_1 > \ldots$ of generic birecurrent train tracks so that*

$$E(\tilde{G}) = \cap_{m \geq 0} E(\tau_m).$$

Proof: It follows directly from Theorem 1.7.12 (using the sequence of train tracks described in the proof of Theorem 1.7.8) that there is a sequence $\tau = \tau'_0 > \tau'_1 > \ldots$, as in the statement of Corollary 1.7.13, but the train tracks τ'_m in fact fail to be generic by construction. To correct this, we take a generic track τ_m arising from τ'_m by a sequence of combs. Evidently,

$$\tau = \tau_0 > \tau'_1 > \tau_1 > \tau'_2 > \ldots,$$

and the assertions about the sequence $\tau > \tau_1 > \ldots$ follow directly from the corresponding facts for the sequence $\tau > \tau'_1 > \ldots$.   q.e.d.

§1.8 BOUNDED SURFACES AND TRACKS WITH STOPS

This section is devoted to sketching the extension of the foregoing theory for surfaces with boundary and then developing a relative version of train tracks (from [P1]) in this setting. Let $\hat{F}_g^{s,r}$ denote a smooth, oriented surface of genus g with $s \geq 0$ distinguished points, whose union we denote Δ, and with $r \geq 1$ smooth boundary components, whose union we denote $\partial\hat{F}_g^{s,r}$. As usual, we will often regard the points of Δ as cusps and define $F_g^{s,r} = \hat{F}_g^{s,r} - \Delta$; when the topological type of the surface $F_g^{s,r}$ (or $\hat{F}_g^{s,r}$, respectively) is fixed or not important, we may call it simply F (or $\hat{F}$).

Define a train track τ in such a surface F exactly as before (and adopt the attendant terminology of branches, switches, etc.), where τ is required to be disjoint from ∂F. The basic fact is that *all* of the results of §1.1 $-$ §1.7 hold in this setting as well. The two small technical distinctions (which contribute to the rationale for presenting the bounded case separately here) are as follows: one must distinguish between two types of complementary n-gon-minus-a-disk depending on whether the smooth frontier edge lies in ∂F; the universal cover of F is not canonically identified with $\mathbb{H}^2$, but rather with a subset of $\mathbb{H}^2$. The former difference is easily handled by simply formally treating each component of ∂F as a puncture. The latter difference is addressed by doubling $F_g^{s,r}$ along all the curves in $\partial F_g^{s,r}$ to produce a surface F' of type F_{2g+r-1}^{2s}; choosing a hyperbolic structure on F', the universal cover of F' can be identified with $\mathbb{H}^2$, and $F \subset F'$ inherits its *intrinsic Poincaré metric*, where each component of $\partial F \subset F'$ is geodesic. Furthermore, a train track in F can be regarded as a train track in $F' \supset F$ in the natural way.

We next briefly discuss some of the particulars of the foregoing theory in this setting. First notice that by Condition (3) in the definition of train track, no curve component of a train track in F can be parallel to a component of ∂F; thus, the arguments of §1.1 apply with only small modifications to show that F contains a train track if and only if $\chi(F) < 0$ and F is not one of the surfaces $F_0^{s,r}$, where $s + r = 3$. A multiple curve in $F = F_g^{s,r}$ is defined exactly as before with the the additional proviso that no component is isotopic into ∂F; a train track $\tau \subset F$ carries multiple curves in F, and

102

a measure on τ uniquely determines a multicurve carried by τ. A pants decomposition of F is a multiple curve $C \subset F$ so that each component of $F - C - \partial F$ is homeomorphic to the interior of a pair of pants, there are $N = 3g - 3 + s + r$ components in a pants decomposition of $F = F_g^{s,r}$, and Dehn's Theorem is proved as before.

Formally treating each component of ∂F as a puncture (so, for instance, if R is an m-gon-minus-a-disk component of $F - \tau$, then the analogue of Move 2 in the proof of Theorem 1.3.6 is applicable in R but Move 3 is not), all of the results of §1.3 apply in this setting. In particular, if $g > 1$ or $r + s > 1$, then any birecurrent train track in $F = F_g^{s,r}$ is a subtrack of a complete track each of whose complementary regions is a trigon, a once-punctured monogon, or a monogon-minus-a-disk whose smooth frontier component lies in ∂F; a birecurrent track in $F_1^{r,s}$, where $r + s = 1$ is a subtrack of a complete track whose unique complementary region is either a once-punctured bigon or a bigon-minus-a-disk where the smooth frontier component lies in ∂F.

Doubling F to produce a punctured surface F' without boundary and regarding a track $\tau \subset F \subset F'$ as a track in F' (as above), transverse recurrence of τ as a track in F implies transverse recurrence of τ as a track in F'. Thus, Theorem 1.4.3 is seen to hold in this setting. Furthermore, regarding τ as a train track in F' leads to the assignment of a point $E(\rho) \in M_\infty$ to a bi-infinite trainpath ρ on $\tilde{\tau}$, where $\tilde{\tau}$ is the full pre-image of τ in the universal cover $\mathbb{H}^2$ of F'; the proof of Theorem 1.5.4 then applies to show that $E(\tau) \subset M_\infty$ is closed.

A lamination L on F is defined as in §1.6 with the further proviso that L is disjoint from ∂F; L is a geodesic lamination if its leaves are geodesic for the intrinsic Poincaré metric. L is said to have compact support if it avoids a neighborhood of $\Delta \cup \partial F$. Arguing in the universal cover $\mathbb{H}^2$ of $F' \supset F$ as in the proofs of Theorems 1.6.5 and 1.6.6, one finds that each geodesic lamination G on F with compact support is carried by some transversely recurrent train track $\tau \subset F$, and, in fact, $G < \tau$ if and only if $E(\tilde{G}) \subset E(\tau)$. A transverse measure on a lamination L on F is defined as before, a measured lamination carried by a track induces a measure on the track, and Construction 1.7.7 applies essentially unchanged to give the continuous two-sided inverse to this assignment. The spaces of (projective) measured laminations on F are defined as before.

A more interesting extension of the theory involves a relative version of train tracks, as follows. Let us choose, once and for all, a collection Σ of points, one point in each component of ∂F, and consider a collection τ of one-dimensional CW complexes disjointly imbedded in F so that $\tau \cap \partial F$ is contained in Σ and consists entirely of vertices of τ. Vertices of τ which lie in $\Sigma \subset \partial F$ are called *stops* of τ, and, as before, vertices of τ which lie in $F - \partial F$ are called *switches*, and edges of τ are called *branches*. Such a CW complex $\tau \subset F$ is called a *train track with stops* in F provided Condition (1) in the definition of train track holds for all switches and stops of τ, where the one-sided tangent vector to each half-branch incident on a stop is required to be transverse to ∂F, Condition (2) holds for all switches of τ, and Condition (3) holds for each component of $F - \Sigma - \tau$; examples of train tracks with stops are given in Figure 1.8.1. A train track τ with stops is *generic* provided that each switch of τ in Σ is either nullvalent or univalent and each switch of τ not in Σ is trivalent (or bivalent if it lies in a simple closed curve component of τ). Notice that a train track with stops may contain a curve component which is homotopic to a component of ∂F (since the double of the corresponding complementary region is a once-punctured torus), but no curve component may be homotopic into a puncture (since the double of the corresponding complementary region would be a twice-punctured-sphere).

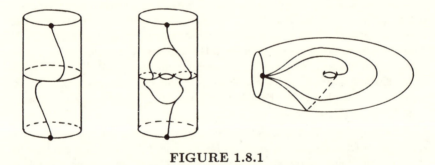

FIGURE 1.8.1

Remark: There are several other possible extensions: for instance, we might allow stops only on certain boundary components, or, indeed, we might allow several different stops on a given component. One might also allow stops at points of Δ, but there are some technical difficulties with the theory of tangential measure (see below) in this case.

The arguments required to prove Proposition 1.1.1 and Corollary 1.1.2 apply without change for train tracks with stops. Suppose that $F = F_g^{s,r}$, where we continue to assume that $r \neq 0$, and consider the double F'' of $F - \Sigma$ along the smooth arcs in $\partial F - \Sigma$; the surface F'' is of type F_{2g+r-1}^{r+2s}, and each boundary component of F gives rise to a puncture of F''. Arguing as before, one finds that F contains a train track with stops if and only if $\chi(F'') < 0$ and F is not the surface $F_0^{1,1}$; we henceforth assume that these conditions hold in addition to assuming (as always) that F is distinct from $F_0^{3,0}$.

Let $\{K_k\}_{k=1}^r$ denote the various components of ∂F, and choose a closed arc $\alpha_k \subset K_k$ which contains $K_k \cap \Sigma$, for each $k = 1, \ldots, r$. A *multiple arc* C in F is a smooth compact one-manifold imbedded in F whose boundary (if any) is contained in $\cup_{k=1}^r \alpha_k$, so that no curve component of C is null-homotopic or homotopic into a puncture, and no arc component of C is homotopic into ∂F. In particular, a multiple curve is a multiple arc, but a multiple arc in F may contain curves which are boundary-parallel, and, of course, may also contain arcs properly imbedded in F. A multiple arc C is *carried* by a train track τ with stops using the same definition as before (where the supporting map is homotopic to the identity by a homotopy which fixes $\partial F - \cup_{k=1}^r \alpha_k$ pointwise), and, if C is carried by τ, then C gives rise to an assignment of a nonnegative integer to each branch of τ, where this assignment satisfies the switch condition at each switch. Conversely, any such assignment of nonnegative integers gives rise to a multiple arc in F as before.

The isotopy class fixing $\partial F - \cup_{k=1}^r \alpha_k$ pointwise of a multiple arc in F is called a *multiarc*, the collection of all multiarcs in F is denoted $\mathcal{R}(F)$, and there is an analogue of Dehn's Theorem for multiarcs, which we next discuss. By a *pants decomposition* of (F, Σ), we mean the union of $\{K_k\}_{k=1}^r$ with a pants decomposition $\{K_i\}_{i=r+1}^{r+N}$ of F (as above), where $N = 3g - 3 + r + s$. Each component of $\{K_i\}_{i=1}^{N+r}$ is called a *pants curve*, and we choose a closed arc $\alpha_i \subset K_i$, for $i = r + 1, \ldots, r + N$. As before, for $i = r + 1, \ldots, r + N$, choose a small closed regular neighborhood A_i of each K_i, which is identified with the standard oriented annulus A by means of a map $\nu_i : A \to A_i$; similarly, for each $k = 1, \ldots, r$, choose a small closed regular neighborhood A_k of K_k in F and an orientation-preserving homeomorphism $\nu_k : S^1 \times [0,1] \subset A \to A_k$, where $\nu_k(S^1 \times \{1\}) = K_k$. Again letting $\mathcal{U}$ denote a small open regular neighborhood of Δ in F, each component P_j of $F - \mathcal{U} - \cup\{\overset{\circ}{A}_i\}_{i=1}^{N+r}$ is a pair of pants imbedded in F, and we may choose an orientation-preserving homeomorphism f_j from the standard oriented pair of pants P to P_j, for $j = 1, \ldots, 2g - 2 + r + s$; as usual, we require that f_j carries each component of $f_j^{-1} \circ \nu_i \circ G^{-1} \circ \nu_i^{-1}(\alpha_i)$ to a window in ∂P whenever $A_i \cap P_j \neq \emptyset$, where G denotes the collapse of A onto its core circle.

Arguing as before, we may isotope a multiple arc into canonical position with respect to a basis (defined as before), define intersection numbers $m_i \geq 0$ and twisting numbers $t_i \in \mathbb{Z}$, for $i = 1, \ldots, N + r$, and arrive at a relative version of Dehn's Theorem which gives a one-to-one correspondence between $\mathcal{R}(F_g^{s,r})$ and the subset of $\mathbb{Z}_+^{N+r} \times \mathbb{Z}^{N+r}$ satisfying Condition (a) and the obvious analogue of Condition (b) in Theorem 1.2.1.

Remark: In analogy to the above, one can also give a parametrization of homotopy classes of appropriate one-manifolds imbedded in $\hat{F} = F \cup \Delta$ with boundary in $\cup_{k=1}^{r} \alpha_k \cup \Delta$ by a certain subset of $\mathbb{Z}_+^{N+r+s} \times \mathbb{Z}^{N+r}$. There is a difference in the treatment of boundary components of $\hat{F}$ and points of Δ since one can isotope an arc incident on a point in Δ so as to get rid of any twisting.

A train track τ with stops is said to be *recurrent* if for each branch b of τ there is a multiple arc C_b carried by τ whose associated measure is positive on b. A *(transverse) measure* on τ is defined as before, and a train track with stops is recurrent if and only if it supports a strictly positive measure. The definition of *hitting efficiently* is exactly the same as before, and Proposition 1.3.3 holds *verbatim*. τ is said to be *transversely recurrent* if for each branch b there is a multiple arc hitting τ efficiently and meeting b, and a track with stops which is both recurrent and transversely recurrent is said to be *birecurrent*. As in Lemma 1.3.3, if τ is a transversely recurrent track with stops, and τ' is a track (perhaps) with stops so that either $\tau \supset \tau'$ or $\tau > \tau'$, then τ' is transversely recurrent (as a track with stops) as well.

Suppose that $\tau \subset F$ is a train track with stops and ν is an assignment of nonnegative real number to each branch of τ. ν is called a *tangential measure* on τ provided that Conditions (1) and (2) hold as before on each component of $F - \tau$ whose frontier is disjoint from ∂F, and ν is said to be *even* if it is $[\mathbb{Z}_+ \cup \{0\}]$-valued and satisfies the evenness condition as before on each such component of $F - \tau$. In anology to Lemma 1.3.4, we claim that if ν is an even tangential measure on $\tau \subset F$, then there is a multiple arc $C \subset F$ hitting τ efficiently so that C intersects each branch b of τ in exactly $\nu(b)$ points. To see this, we may add curve components to τ to produce a train track with stops so that each complementary region is one of the following: *i*) an m-gon, where $m \geq 3$; *ii*) a once-punctured-m-gon, where $m \geq 1$; *iii*) an m-gon-minus-a-disk, where $m \geq 1$; *iv*) a pseudo pair of pants; *v*) an annulus with one frontier edge in ∂F. The smooth frontier curves in cases *iii*) and *iv*) may be taken to be disjoint from ∂F, we may assume that each component of type *ii*) or *iii*) satisfies $m = 1$, as before, and one easily produces an even tangential measure ν'' on the resulting train track τ'' with stops which extends the tangential measure ν on τ. Choose $\nu''(b)$ distinct points on each branch b of τ'' and connect these points exactly as before in each component of $F - \tau''$ whose frontier is disjoint from ∂F. Since a half-branch of τ'' incident on a stop meets ∂F transversely, a component of $F - \tau''$ whose frontier meets ∂F must be of type *i*) or type *v*). If R is a component of the latter sort whose frontier edge in $F - \partial F$ has total ν''-tangential measure ξ, simply choose ξ points in the other frontier edge of R, and then connect the two frontier edges of R with ξ arcs properly imbedded in R as in Figure 1.8.2a. Consider an m-gon component of the former sort, and proceed by induction on $m \geq 3$. For the basis step $m = 3$, notice that exactly one frontier edge, say δ, of R lies in ∂F, let ξ_1 and ξ_2, respectively, denote the total ν''-tangential measures of the other frontier edges, say δ_1 and δ_2, and let ξ denote the minimum of ξ_1 and ξ_2. Connect the ξ points on δ_1 furthest from δ to the corresponding points on δ_2, choose $|\xi_1 - \xi_2|$ points on δ, and finally connect the remaining points on $\delta_1 \cup \delta_2$ to the chosen points on δ; see Figure 1.8.2b. For the inductive step, suppose that R is an m-gon of the former sort, add a branch b to τ'' and extend ν'' to an even tangential measure on the resulting track with stops, as before, to decompose R into a trigon and an $(m-1)$-gon; add arcs in each of these regions using the inductive hypothesis to finally construct the required multiple arc.

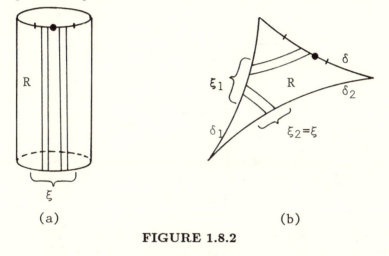

(a)   (b)

FIGURE 1.8.2

Remark: Notice that if we were to allow stops at points of Δ, then the construction above might produce puncture-parallel curves in the one-submanifold associated to an even tangential measure on a track with stops.

Arguing as in the proof of Corollary 1.3.5, one concludes that a train track τ with stops in F is transversely recurrent if and only if it supports a strictly positive even tangential measure, and that this, in turn, is equivalent to the condition that there is a multiple arc in F which hits τ efficiently and meets every branch of τ.

A birecurrent train track with stops is said to be *complete* if it is not a proper subtrack of any birecurrent track, and we have the following analogue of Theorem 1.3.6: if $g > 1$, $s > 1$, or $r \geq 1$, then any birecurrent train track with stops on $F_g^{s,r}$ is a subtrack of a complete train track with stops, each of whose complementary regions is either a trigon or a once-

punctured monogon; any birecurrent train track with stops on $F_1^{1,0}$ is a
subtrack of a complete train track whose unique complementary region is
a once-punctured bigon. To prove this result, in addition to the Moves 1-6
employed before (where the added branches may be incident on stops), we
must also make use of the

> **Move:** Suppose that R is an annulus complementary region one of
> whose frontier edges lies in ∂F. Add a branch in R connecting a
> switch in the other frontier edge of R to the stop in the frontier of
> R; see Figure 1.8.3.

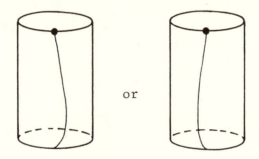

FIGURE 1.8.3

To see that the new moves preserve transverse recurrence, one makes use
of parallel translates of the various boundary components of F.

Orientability of a train track with stops is defined exactly as before,
and Proposition 1.3.7 is seen to hold in this setting. The extension to a
complete train track with stops is pursued as follows: one constructs a
graph G with a solid vertex for each component of the track and with a
hollow vertex for each complementary region which is an m-gon-minus-a-
disk, a pseudo pair of pants, or an annulus with one frontier edge in ∂F,
and adjoins edges to G, as before; a hollow univalent vertex corresponds to
either a twice-punctured-nullgon or perhaps an annulus with one frontier
edge in ∂F. Reduce to the case that G is a tree, then further reduce to
the case that G is a single solid vertex by using the various moves desribed
above, and finally finish the argument just as before.

A train track with stops is said to be *generic* provided that each switch
is (either bivalent) or trivalent, and each stop is either nullvalent or univalent. One can *comb* a non-generic track with stops near switches (as in
Figure 1.4.2) and near stops (as in Figure 1.8.4) to produce a generic track
with stops. The technique of *sneaking up* is applicable to generic train tracks

with stops, and the analogues of Proposition 1.4.1 and Corollary 1.4.2 are proved as before.

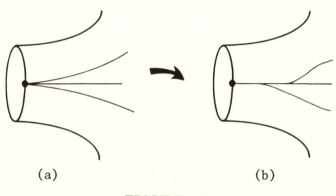

(a)   (b)

FIGURE 1.8.4

Before further discussing the extensions of §1.4 − §1.7 in our current context, we must describe our point of view on metrics. Recall the surface F'' discussed above which arises from F by doubling along all of the smooth arcs of $\partial F - \Sigma$. By our assumption that $\chi(F'') = 4 - 4g - 3r - 2s < 0$, F'' supports a hyperbolic structure, and we have furthermore assumed that $r \neq 0$, so that F'' is connected; choosing such a structure on F'', the universal cover of F'' may be identified with $\mathbb{H}^2$. Regard $F - \Sigma$ as a subset of F'' in the natural way, so that a component of $\partial F - \Sigma$ is identified with a bi-infinite geodesic in F'' which runs from a puncture of F'' back to itself. Since we furthermore assume that $F \neq F_0^{1,1}$, given a component ∂ of ∂F and the corresponding component $\partial^* = \partial - \Sigma$ of $\partial F - \Sigma$, there is a unique simple closed geodesic $\partial_* \subset F - \Sigma \subset F''$ which represents the homotopy class of ∂, and $\partial^* \cup \partial_*$ bounds a monogon-minus-a-disk in F''. Each such region is called a *spike*, and each component of $\partial F - \Sigma$ gives rise to a distinct spike (even though it may happen that $\partial_* = \partial'_*$ for distinct components $\partial \neq \partial'$ of ∂F when $F = F_0^{0,2}$); see Figure 1.8.5 for some examples. The inherited metric on $F - \Sigma \subset F''$ is called an *intrinsic Poincaré metric with spikes*. Notice that if $\chi(F) < 0$, then removing the collection of all spikes from $F - \Sigma \subset F''$ produces an imbedded copy of F in F'', and the hyperbolic structure on F'' restricts to an intrinsic Poincaré metric on $F - \Sigma - \{\text{spikes}\}$.

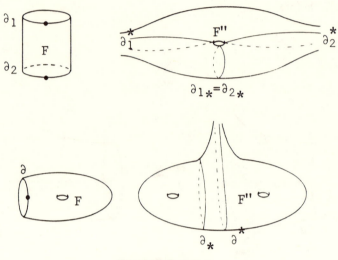

FIGURE 1.8.5

If $\tau \subset F$ is a generic train track with stops and $F - \Sigma$ is given its intrinsic Poincaré metric with spikes, then each half-branch of τ which is incident on a stop of τ is to be regarded as a semi-infinite path in $F - \Sigma$ which is asymptotic to the stop. Furthermore, if τ is a complete generic track with stops, then for each stop ς, there is a unique branch b_ς of τ which is incident on ς, and the trigon component R_ς of $F - \tau$ which contains b_ς in its frontier is to be regarded as an "ideal trigon" in $F - \Sigma - \tau$, each of whose frontier edges is bi-infinite; see Figure 1.8.6.

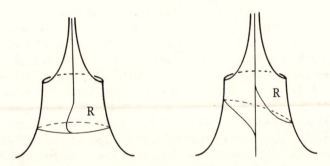

FIGURE 1.8.6

The analogue of Theorem 1.4.3 in our current setting posits the equivalence of the following conditions on a generic train track $\tau \subset F$ with stops: (i) τ is transversely recurrent; (ii) τ supports a positive tangential measure satisfying strict inequalities in Conditions (1) and (2) in the definition of tangential measure on each component of $F - \tau$ whose frontier is disjoint from ∂F; (iii) for every $\varepsilon > 0$ and $L > 0$, there is an intrinsic Poincaré metric with spikes on F so that τ has geodesic curvature less than ε at each point, and each branch of τ has length at least L. The proof of the equivalence of Conditions (i) and (ii) above is analogous to the proof given before. To see that Condition (ii) implies Condition (iii) above, simply apply the arguments given before to $F - \cup_{\varsigma \in \Sigma} R_\varsigma$ and then adjoin a collection of ideal triangles, one triangle for each $\varsigma \in \Sigma$. Finally, to see that Condition (iii) implies condition (ii), choose an intrinsic Poincaré metric with spikes so that τ has small curvature and long branches, and let $\nu(b)$ denote the length of b in this structure for each branch b which is not incident on a stop. As before, ν is seen to be a tangential measure on τ, and we may extend ν to the required tangential measure on τ with an essentially arbitrary assignment of tangential measure to each branch which is incident on a stop (if any).

   As in Proposition 1.4.9, one may perform trivial collapses on admissible arcs to produce a complete train track with stops from a birecurrent one after first adding branches incident on stops (if necessary) as in Figure 1.8.3.

   The analogues of the results of §1.5 in our current setting are easily derived, as follows. Suppose that $\tau \subset F$ is a transversely recurrent train track with stops, regard $\tau \subset F - \Sigma \subset F''$ as above, and consider the full pre-image $\tilde{\tau}$ of τ in the universal cover $\mathbb{H}^2$ of F''. Suppose that ρ is a trainpath on $\tilde{\tau}$ of one of the following types: ρ is a bi-infinite trainpath; ρ is a semi-infinite trainpath that begins at a point of S^1_∞ which corresponds to a stop of τ; ρ is a finite trainpath on $\tilde{\tau}$ that connects two points of S^1_∞ which correspond to stops of τ. In each of these cases, there is a corresponding point $E(\rho) \in M_\infty$, and, up to reversal of trainpaths, $E(\rho)$ uniquely determines ρ, as before. Defining

$$E(\tau) = \{ E(\rho) : \rho \text{ is a trainpath as above on } \tilde{\tau} \},$$

one proves as before that $E(\tau) \subset M_\infty$ is closed. Finally, if $\tau_1, \tau_2 \subset F$ are transversely recurrent train tracks with stops and $\tau_1 < \tau_2$, then $E(\tau_1) \subset E(\tau_2)$.

   A *lamination* L in $F - \Sigma$ is defined as before, and L is a *geodesic lamination* if it is geodesic for some fixed intrinsic Poincaré metric with spikes. The analogues of Corollary 1.6.2 (where each element of a corresponding $\pi_1(F'')$-invariant subset is required to span a geodesic which lies in the full pre-image of $F \subset F''$ in $\mathbb{H}^2$) and Proposition 1.6.3 are proved as before. Let $\mathcal{U}$ be the deleted unit horoball neighborhood of the collection of punctures of F; a geodesic lamination in $F - \Sigma$ is said to have *compact support* if it is disjoint from some neighborhood of the punctures of $F - \Sigma$, and Proposition 1.6.4 holds as before. As an addendum to this result, we next prove that if a half-leaf of a geodesic lamination meets a spike, then it must be asymptotic to the corresponding stop. To see this, suppose that g is a leaf of the geodesic lamination G in F and g meets a spike corresponding to the stop ς, say with frontier edges ∂_* and ∂^*, where ∂_* is a closed geodesic and ∂^* is a bi-infinite geodesic running from the puncture ς of F'' back to itself. Of course, $g \cap \partial^* = \emptyset$, and we may choose a parametrization $\hat{g}: [0, \infty] \to F$ of a half-leaf $\hat{g}$ of g by arc-length so that $\hat{g}(0) \in \partial_*$ and the tangent $\hat{g}'(0)$ points into the spike. It follows that $\hat{g}(t) \notin \partial_*$ for all $t > 0$, since otherwise, there is a bigon in F with geodesic boundary (contained in $\partial_* \cup \hat{g}$), which is absurd. Furthermore, if $\hat{g}$ meets a small horocycle about the puncture ς of F'', then the argument given before shows that $\hat{g}$ tends to ς. Thus, if $\hat{g}$ is not asymptotic to ς, then $\hat{g}$ is constrained to lie inside the spike and outside a horoball about ς. By compactness of (the unit tangent bundle of) this region, we can extract from $\{\hat{g}(m) : m \in \mathbb{Z}_+\}$ a subsequence, say $\{\hat{g}(m_k) : k \in \mathbb{Z}_+\}$ so that both $\hat{g}(m_k)$ and the tangents $\hat{g}'(m_k)$ converge. Connecting $\hat{g}(m_k)$ and $\hat{g}(m_{k+1})$ by a geodesic arc, we thus produce a piecewise geodesic curve α_k, for $k \in \mathbb{Z}_+$, so that the exterior angles at the two non-smooth points of α_k tend to π. Since $\hat{g}$ is simple and $\hat{g} \cap \partial_* \neq \emptyset$, we cannot have α_k tending to ∂_*, so our assumptions lead to a contradiction of the Gauss-Bonnet Theorem. This completes the proof of our addendum to Proposition 1.6.4.

In analogy to Theorem 1.6.5, we claim that every geodesic lamination in $F - \Sigma$ with compact support is carried by some transversely recurrent train track with stops in F. To prove this, one argues as before in the compactum $F - \mathcal{U} - \{\text{spikes}\}$ and uses the addendum to Proposition 1.6.4 given above. Finally, arguing in analogy to the proof of Theorem 1.6.6, one finds that a transversely recurrent train track $\tau \subset F$ with stops carries a geodesic lamination G in $F - \Sigma$ with compact support if and only if $E(\tilde{G}) \subset E(\tau)$.

A *transverse measure* λ on a (geodesic) lamination L in $F - \Sigma$ is defined as before, and the pair (L, λ) is called a *measured (geodesic) lamination*. The simplest example of a measured geodesic lamination (G, λ) in $F - \Sigma$ is when G corresponds to a multiple arc in F, each of whose components is geodesic, and λ arises from a weight in $\mathbb{R}_+$ assigned to each component of G. The analogue of Proposition 1.7.1 is proved as before.

The analogue of Proposition 1.7.2 is as follows. If g is a leaf in a measured geodesic lamination G in F of compact support, then exactly one of the following possibilities holds: g is a closed geodesic curve which is isolated from $G - g$ and disjoint from the spikes; g is a geodesic arc which is asymptotic to exactly two (not necessarily distinct) stops, and g is isolated from $G - g$; g is a bi-infinite geodesic which is disjoint from the spikes. (In particular, no leaf can be asymptotic to exactly one stop.) To see this, we argue as before to conclude that every simple closed geodesic leaf of G is isolated . Furthermore, arguing as before, any isolated half-leaf of G must meet a spike, so by the addendum to Proposition 1.6.4 above, such a half-leaf must be asymptotic to a stop; it follows that an isolated bi-infinite leaf of G must be asymptotic to stops in both directions. To finish the proof, it remains only to show that a leaf of G which connects two stops is necessarily isolated. To this end, no two such leaves can be properly homotopic (since they are geodesic), and, since they are pairwise disjoint, there can be only finitely many such leaves. Thus, for any such leaf g, the intersection of g with a spike must be isolated, and one again concludes that g is actually isolated in F.

It follows that a measured geodesic lamination in $F - \Sigma$ with compact support decomposes uniquely into a finite number of closed and isolated minimal sets, where each minimal set is a geodesic curve with atomic transverse measure, a geodesic arc connecting stops with atomic transverse measure, or a measured geodesic lamination G each of whose leaves is a bi-infinite geodesic which is dense in G and disjoint from the spikes. Furthermore, if G is a measured geodesic lamination in $F - \Sigma$ with compact support and $\alpha \in \Lambda(G)$, then $\alpha \cap G$ is the disjoint union of a finite set and a Cantor set, and the isolated points of $\alpha \cap G$ are exactly the intersections of α with the geodesic curves and geodesic arcs in G connecting stops. Proposition 1.7.5 holds exactly as before, each measured geodesic lamination in $F - \Sigma$ with compact support is carried by some birecurrent train track with stops, and if G is a measured geodesic lamination in $F - \Sigma$ which is carried by the train track $\tau \subset F$ with stops, then G induces a well-defined measure on τ.

The analogue of Construction 1.7.7 is pursued just as before, where the vertical arcs of rectangles corresponding to ends of branches incident on stops are simply removed from the bi-foliated neighborhood, and the lamination L is derived from the leaves of the bi-foliated neighborhood. We claim that no leaf of L can be asymptotic to exactly one stop, and the proof is as follows. If b is a branch which is incident on a stop ς, then let ξ be a tie in the rectangle associated to b. Since a singular leaf can meet b at most once, the set $\theta = \xi \cap \{\text{singular leaves}\}$ is finite, and the closed set $K \subset \eta$ (constructed in Step 1) consists of a finite collection of closed intervals. It follows that if ℓ is a leaf of L which is asymptotic to ς, then there is a packet of leaves of L which is parallel to L corresponding to such an interval in K. If ℓ is not asymptotic to stops at both ends, we find a packet of leaves which has infinite area (where the area of a subset S of the bi-foliated neighborhood is defined to be the sum over all the rectangles R in Step 1 of the Euclidean areas of $S \cap R$) imbedded in the bi-foliated neighborhood; since the bi-foliated neighborhood itself has finite area, this is a contradiction, and our claim is proved. L is replaced by the corresponding geodesic lamination G in $F - \Sigma$ and a transverse measure on G is derived as before.

This analogue of Construction 1.7.7 is seen to give rise to a continuous injection from the space of transverse measures on a birecurrent train track with stops in F to the space $\mathcal{ML}(F)$ whose image consists of the collection of measured geodesic laminations in $F - \Sigma$ with compact support that are carried by the track.

CHAPTER 2   COMBINATORIAL EQUIVALENCE

An equivalence relation on transversely measured train tracks is described which is generated by isotopy and two combinatorial moves called splitting and shifting. Equivalent measured tracks give rise to the same measured geodesic lamination (assuming the tracks are transversely recurrent), and the converse of this result is proved later in this chapter. There is a natural $\mathbb{R}_+$-action on the set of measures supported on a track, and the resulting classes are called projective measures on the track. The collection of all projective measures on a fixed train track is found to have the natural structure of a polyhedron whose faces correspond exactly to the recurrent subtracks. We then turn to certain useful facts, as follows. The equivalence relation generated by splitting and isotopy alone is shown to coincide with the one considered above. It is easy to see that if a track splits and shifts to another track, then the former track carries the latter one, and a converse to this result is given. Given a fixed multiple curve, we show how to split a measured train track until it hits the multiple curve efficiently. We finally introduce standard models for measured train tracks (from [P1]) proving that any measured train track is equivalent to a unique standard model. These standard models are used to prove that if two measured train tracks give rise to the same measured geodesic lamination, then the two train tracks are equivalent. The construction of a measured geodesic lamination from a measured train track is extended to the setting where the train track is not necessarily transversely recurrent.

§2.1 SPLITTING, SHIFTING, AND CARRYING

For the duration of this chapter, the train tracks we consider will be assumed to be generic and recurrent unless otherwise stated. We will not, however, assume that train tracks are transversely recurrent unless this is explicitly stated. Furthermore, for simplicity, we will restrict attention to train tracks in a surface $F = F_g^s$; the extensions of our results to bounded surfaces and to train tracks with stops are straight-forward and left to the reader.

Given a train track $\tau \subset F$, define the collection $V(\tau)$ of all (not necessarily non-zero) transverse measures supported on τ. Let n denote the number of branches and m the number of switches of τ, and enumerate the weights of branches of τ to get an imbedding of $V(\tau)$ into $[\mathbb{R}_+ \cup \{0\}]^n$. The collection of switch conditions imposes a family of m (not necessarily independent) linear constraints on the image. It follows that there is a natural identification between $V(\tau)$ and the intersection of $[\mathbb{R}_+ \cup \{0\}]^n$ with a linear subspace of $\mathbb{R}^n$. Define the *dimension d* of τ to be the dimension of this corresponding linear subspace. Of course, we have the inequality $d \geq n - m$, and there is moreover the following result.

Lemma 2.1.1: *If $\sigma \subset F$ is a connected, recurrent, non-orientable track of dimension d with n branches and m switches, then the dimension of σ is $d = n - m$.*

Proof: We must show that the switch conditions are linearly independent on σ. Suppose to the contrary that $\hat{b}$ is a half-branch incident on a switch v of σ and the switch condition corresponding to v lies in the span of the other switch conditions on σ. By Basic Fact 2 in Proposition 1.3.7, there is a trainpath ρ on σ beginning and ending at v which reverses direction along $\hat{b}$. Consider the function μ on the set of branches of σ defined by taking $\mu(e)$ to be the number of times ρ traverses the branch e of σ. By construction, this function μ satisfies the switch condition at each switch of σ other than v and fails to satisfy the switch condition at v, leading to the desired contradiction.   q.e.d.

In light of remarks above, for any recurrent train track τ of dimension d, $\mu \in V(\tau)$ if and only if $t\mu \in V(\tau)$ for each $t \in \mathbb{R}_+$, and $V(\tau)$ is therefore identified with the cone from the origin of a closed cell $U(\tau)$ of dimension $d - 1$. An element of $U(\tau)$ (i.e., the class of a non-zero measure on τ under the natural $\mathbb{R}_+$-action on $V(\tau)$) is called a *projective measure* on τ. If $\mu \in V(\tau) - \{\vec{0}\}$, then the corresponding projective measure will be denoted $\bar{\mu} \in U(\tau)$, and we say that $\bar{\mu}$ is *positive* if some (hence any) $\mu \in \bar{\mu}$ satisfies $\mu > 0$.

If $\bar{\mu} \in U(\tau)$, then choose $\mu \in V(\tau) - \{\vec{0}\}$ with $\mu \in \bar{\mu}$, and consider the collection $B_{\bar{\mu}}$ of branches on which μ vanishes. Define the subtrack $\tau_\mu = \tau_{\bar{\mu}} = \tau - B_{\bar{\mu}} \subset \tau$ (where we amalgamate each appropriate resulting bivalent switch), and notice that the measure μ on τ gives rise to a *positive* projective measure (which will also be denoted $\bar{\mu}$) on $\tau_{\bar{\mu}}$. It follows from Proposition 1.3.1 that $\tau_{\bar{\mu}}$ is recurrent, so the projective measure $\bar{\mu}$ on τ canonically gives rise to a recurrent subtrack $\tau_{\bar{\mu}} \subset \tau$.

Given a non-empty recurrent subtrack $\sigma \subset \tau$, define the associated *face* of $U(\tau)$ to be

$$\{\bar{\mu} \in U(\tau) : \tau_{\bar{\mu}} \subset \sigma\}.$$

A projective measure on σ corresponds to a projective measure on τ lying in the face of $U(\tau)$ associated to σ in the natural way, and we henceforth identify the closed cell $U(\sigma)$ with the face of $U(\tau)$ associated to σ. In the same way, we identify $V(\sigma)$ with the corresponding *face*

$$\{\mu \in V(\tau) : \tau_\mu \subset \sigma\}$$

of $V(\tau)$.

Lemma 2.1.2: *If τ is a birecurrent train track of dimension d, then the faces of $U(\tau)$ give it the structure of a polyhedron of dimension $d - 1$.*

Proof: As remarked above, each face of $U(\tau)$ is a closed cell, and, by definition, the interiors of any two distinct faces are disjoint. To prove the lemma, we must furthermore show that the intersection $U(\sigma_1) \cap U(\sigma_2)$ of two faces $U(\sigma_1)$ and $U(\sigma_2)$ of $U(\tau)$ is either empty or again a face of $U(\tau)$. Letting σ denote the maximal recurrent subtrack of $\sigma_1 \cap \sigma_2$, the equality

$$U(\sigma_1) \cap U(\sigma_2) = U(\sigma)$$

is easily checked (where $U(\emptyset) = \emptyset$ by convention), and the result follows. q.e.d.

For any recurrent track τ, we refer to $U(\tau)$ with its associated polyhedral structure as the *polyhedron of projective measures* on τ and to $V(\tau)$

with its associated structure as the cone on a polyhedron as the *cone of measures* on τ.

Suppose that $\tau \subset F$ is a train track. If $\hat{b}$ is a half-branch of τ and v is a trivalent switch of τ on which $\hat{b}$ is incident, then we say that $\hat{b}$ is *large* if any smooth open arc in τ through v intersects $\hat{b}$ (as in Figure 2.1.1a); the half-branch $\hat{b}$ is said to be *small* if it is not large (as in Figure 2.1.1b). These attributes of small and large descend to ends of branches, and we say that a branch itself is large (or small) if each of its ends is large (small, respectively). Notice that a large branch is necessarily incident on two distinct switches. Similarly, we may call a branch a *mixed* if exactly one end of it is large; notice that recurrence of τ implies that a mixed branch is necessarily incident on two distinct switches. More generally, a recurrent track can contain no closed trainpath consisting entirely of mixed branches.

large   small

(a)   (b)

FIGURE 2.1.1

If e is a mixed branch of τ, then we may *shift* τ along e by performing an inverse comb (collapsing e to a point) followed by a comb (which generates a new branch e') to produce a generic track τ', as in Figure 2.1.2a. Notice that if we shift τ' along e', then we simply recover the train track τ. Adopt the notation indicated in Figure 2.1.2a for the branches of τ which are incident on the switches in the closure of e; there is a natural one-to-one correspondence between the set of branches of τ and those of τ', and if f is a branch of τ, then we let f' denote the corresponding branch of τ' as in Figure 2.1.2a. If μ is a (positive) measure on τ, then there is a corresponding (positive) measure μ' on τ' defined by

$$\mu'(f') = \begin{cases} \mu(b) + \mu(c), & \text{if } f' = e'; \\ \mu(f), & \text{otherwise.} \end{cases}$$

In particular, recurrence of τ implies recurrence of τ' by Proposition 1.3.1.

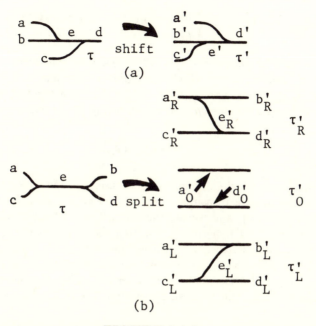

FIGURE 2.1.2

If e is a large branch of τ, then we may *split* τ along e to produce the three tracks τ'_R, τ'_O, and τ'_L indicated in Figure 2.1.2b; not all of these tracks are necessarily recurrent. (See Lemma 2.1.3 below.) The first possibility is called a *right split*, the second a *collision*, and the third a *left split along* e, respectively. The inverse of a split is called a *collapse*, and the inverse of a collision is called a *trivial collapse* (cf. Proposition 1.4.9). Notice that $\tau'_O \subset \tau'_R$ and $\tau'_O \subset \tau'_L$.

If μ is a positive measure on τ and e is a large branch of τ, then there is an associated positively measured track defined as follows. Adopt the notation indicated in Figure 2.1.2b for the branches of τ which are incident on the endpoints of e. There are three cases to consider: $\mu(a) > \mu(b)$ (so $\mu(d) > \mu(c)$ by the switch conditions), and we perform a right split on e; $\mu(a) = \mu(b)$ (so $\mu(c) = \mu(d)$), and we perform a collision along e; $\mu(a) < \mu(b)$ (so $\mu(d) < \mu(c)$), and we perform a left split along e. Formally introducing the variable $X \in \{R, L\}$ and supposing that τ'_X is determined by the measure μ as above, observe that there is a natural one-to-one correspondence between the set of branches of τ and those of τ'_X; if f is a branch of τ, then let f'_X denote the corresponding branch of τ'_X as indicated in Figure 2.1.2b. There is a measure μ'_X induced on τ'_X according to the rule

$$\mu'_X(f'_X) = \begin{cases} |\mu(a) - \mu(b)|, & \text{if } f'_X = e'_X; \\ \mu(f), & \text{otherwise.} \end{cases}$$

Of course, two branches of τ typically amalgamate to form each of the branches a'_O and d'_O of τ'_O which are indicated in Figure 2.1.2b. To any branch f'_O of τ'_O other than a'_O or d'_O, there is a corresponding branch f of τ, and μ induces a measure μ'_O on τ'_O according to the rule

$$\mu'_O(f'_O) = \mu(f),$$

if f'_O is a branch of τ'_O. Thus, in each case, the measured track (τ, μ) uniquely determines a measured track $(\tau', \mu') = (\tau'_X, \mu'_X)$, for the appropriate $X \in \{R, O, L\}$, and we say that (τ', μ') arises from (τ, μ) by a *split along* the branch e. Furthermore, inverting the formulas above, if τ arises from τ' by a collapse, then a (positive) measure μ' on τ' gives rise to a (positive) measure μ on τ; and we say that the measured track (τ, μ) arises from (τ', μ') by a *collapse*. As before, we remark that recurrence is invariant under both splitting and collapsing measured tracks.

Lemma 2.1.3: *Suppose that τ is a recurrent track and let τ'_R, τ'_O, and τ'_L, respectively, be the tracks produced from a right split, a collision, and a left split along some large branch. Either all three tracks τ'_R, τ'_O, and τ'_L are recurrent, or exactly one of them is recurrent.*

Proof: Since τ is recurrent, it supports a positive measure μ, and (τ, μ) splits to the measured track (τ'_X, μ'_X), for some $X \in \{R, O, L\}$, with $\mu'_X > 0$; thus, at least one of the tracks τ'_R, τ'_O, and τ'_L is recurrent. To complete the proof, we must show that recurrence of two tracks among τ'_R, τ'_O, and τ'_L implies recurrence of the third. Since a train track is recurrent if and only if each connected component is recurrent, we may assume that τ (and hence also τ'_R and τ'_L) is connected. Finally, we adopt the notation of Figure 2.1.2b for the branches of τ, τ'_R, τ'_O, and τ'_L.

We begin by showing that recurrence of τ'_R and τ'_O implies recurrence of τ'_L, and, insofar as $\tau'_O \subset \tau'_L$ is recurrent, τ'_O lies in the maximal recurrent subtrack of τ'_L; to prove the implication, we must therefore produce a closed trainpath on τ'_L which traverses the branch e'_L. If τ'_O is connected and nonorientable, then by Basic Fact 2 of Proposition 1.3.7, there is a trainpath on τ'_O which begins by traversing a'_O and ends by traversing d'_O with the respective orientations indicated in Figure 2.1.3a. This trainpath gives rise to a corresponding trainpath on $\tau'_L \supset \tau'_O$, and it is an easy matter to construct the required closed trainpath on τ'_L.

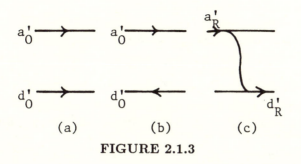

FIGURE 2.1.3

If τ'_O is connected and orientable, then there are two cases depending on the relative orientations of a'_O and d'_O. If these orientations "agree" as in Figure 2.1.3a (or if both orientations are reversed), then by Basic Fact 1 of Proposition 1.3.7, there is a trainpath on τ'_O which begins by traversing a'_O and ends by traversing d'_O (in their indicated orientations), and one produces the required closed trainpath on τ'_L as before. If the orientations on a'_O and d'_O "disagree" as in Figure 2.1.3b (or if both orientations are reversed), then it follows that τ'_R is non-orientable. By Basic Fact 2 of Proposition 1.3.7, τ'_R must support a trainpath which begins and ends by traversing e'_R with opposite orientations, and this is clearly impossible; thus, this second case is not tenable.

To complete the proof of the implication, we must finally consider the possibility that τ'_O is disconnected. In this case, recurrence of τ'_R implies that each component of τ'_O is non-orientable. Orient a'_O and d'_O as indicated in Figure 2.1.3b, apply Basic Fact 2 of Proposition 1.3.7 to produce a trainpath on τ'_O which begins by traversing a'_O (in its prescribed orientation) and ends by traversing the reverse of a'_O, and construct an analogous trainpath on τ'_O traversing d'_O. These trainpaths give rise to corresponding trainpaths on $\tau'_L \supset \tau'_O$, and these combine with two trainpaths of length one on τ'_L which traverse e'_L (in opposite directions) in the natural way to produce the required closed trainpath on τ'_L.

Arguing in analogy to the above, one finds that recurrence of τ'_O and τ'_L similarly implies recurrence of τ'_R, and it remains only to show that if τ'_R and τ'_L are recurrent, then τ'_O is recurrent as well. If τ'_O is connected, then it is evident that at most one of τ'_R and τ'_L is orientable, and we may suppose that τ'_R is non-orientable. Give a'_R and d'_R the respective orientations indicated in Figure 2.1.3c. By Basic Fact 2 of Proposition 1.3.7, there is a trainpath on τ'_R which begins by traversing d'_R and ends by traversing a'_R (in their specified orientations), and there is some sub-trainpath $\rho_R \subset \tau'_O \subset \tau'_R$ with these same properties. Fix a branch f'_O of τ'_O, and consider the branch f'_L of τ'_L containing it; by recurrence of τ'_L, there is some closed trainpath ρ on τ'_L traversing f'_L. Identifying $\tau'_O \subset \tau'_R$ with $\tau'_O \subset \tau'_L$ in the natural way, one produces a closed trainpath on τ'_O traversing f'_O by replacing each appearance of e'_L in ρ by the path $\rho_R \subset \tau'_O$. Thus, f'_O lies in the maximal recurrent subtrack of τ'_O; since f'_O was arbitrary, recurrence of τ'_O is assured, and the proof is complete under the assumption that τ'_O is connected.

Finally, suppose that τ'_O is disconnected and each of τ'_R, τ'_L is recurrent. Let us concentrate on the component τ_+ of τ'_O containing a'_O. Since τ'_R is recurrent and τ'_O is disconnected, there is some trainpath ρ_R on τ'_R which is disjoint from e'_R and begins and ends by traversing a'_R with opposite orientations. Identify $\tau'_O \subset \tau'_R$ with $\tau'_O \subset \tau'_L$ as before, and suppose that $f'_O \subset \tau_+$ is a branch of τ'_O contained in the branch f'_L of τ'_L. Since τ'_L is recurrent, there is some closed trainpath ρ on τ'_L traversing f'_L, and since τ'_O is disconnected, ρ decomposes into the concatenation of trainpaths which either lie entirely in τ_+ or lie in $\tau_L - \tau_+$ and begin and end by traversing e'_L with opposite orientations. One then replaces each sub-trainpath of the latter type by a copy of the trainpath ρ_R to produce a closed trainpath in τ_+ traversing f'_O. Since f'_O was arbitrary, τ_+ is recurrent, as desired. An analogous argument handles the component $\tau'_O - \tau_+$ of τ'_O, and the proof is complete.   q.e.d.

Remark: Some examples to show that that the latter possibilities in the previous lemma actually occur are given in Figure 2.1.4: in Figure 2.1.4a, only a collision along e yields a recurrent track, and in Figure 2.1.4b, only a left split along e yields a recurrent track.

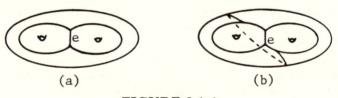

(a)   (b)

FIGURE 2.1.4

We say that two measured train tracks (τ_1, μ_1) and (τ_2, μ_2) are *equivalent* if the associated positively measured tracks $(\tau_{1\mu_1}, \mu_1)$ and $(\tau_{2\mu_2}, \mu_2)$ are related by a composition of shifts, splits, collapses, and isotopies. We will usually regard isotopic tracks as identical. The equivalence class of a measured track (τ, μ) will be denoted $[\tau, \mu]$. If (τ_2, μ_2), where $\mu_2 > 0$, arises from (τ_1, μ_1), where $\mu_1 > 0$, by a composition of shifts, splits, and isotopies (but no collapses), then we say that (τ_1, μ_1) *refines* to (τ_2, μ_2). We may also say simply that τ_1 *refines* to τ_2 (with no mention of measures on the tracks) if there are measures μ_1, μ_2 as above. It follows from remarks above that if τ_1 refines to τ_2, then $\tau_2 < \tau_1$, and, furthermore, τ_1 is recurrent if and only if τ_2 is recurrent.

To close this section, we investigate the relationship between $V(\tau')$ and $V(\tau)$ in case $\tau' < \tau$ and let ϕ denote a supporting map for this carrying. We define the corresponding *incidence matrix* M^ϕ as follows. Enumerate the branches of τ and τ' as $\{b_i\}_1^n$ and $\{b'_i\}_1^{n'}$, respectively. Let $\alpha_i \subset F$ be an arc meeting τ in a single point of b_i which is a regular point of ϕ, for each $i = 1, \ldots, n$, and assume that the arcs α_i are pairwise disjoint. Define the n'-by-n matrix $M^\phi = (M_{ij}^\phi)$ of the supporting map ϕ by letting M_{ij}^ϕ be the number of times $\phi^{-1}(\alpha_i)$ meets b'_j, for $i = 1, \ldots, n$ and $j = 1, \ldots, n'$.

The relevance of the incidence matrix is as follows. If (G, λ) is a measured geodesic lamination with $G < \tau'$, say with supporting map ψ', then there is an induced measure μ' on τ' as discussed after Corollary 1.7.4. The composition $\psi = \phi \circ \psi'$ gives a supporting map for the carrying $G < \tau$, and, by definition, the supporting map ψ induces the measure $M^\phi \mu'$ on τ (where we regard a measure on τ or τ' as a tuple of weights on branches). We therefore think of M^ϕ as a map

$$M^\phi : V(\tau') \to V(\tau)$$

which describes the transformation of induced measures corresponding to measured geodesic laminations carried by τ'.

Remark: In fact, we shall see in Theorem 2.7.4 that $V(\tau)$ is identified with the collection of measured geodesic laminations carried by τ even in our current setting of not necessarily transversely recurrent train tracks. Identifying these two sets, we see that the incidence matrix describes the inclusion $V(\tau') \subset V(\tau)$ in case $\tau' < \tau$. See Proposition 2.2.4 below for a discussion of these matters in case τ is transversely recurrent.

We emphasize that the incidence matrix M^ϕ depends on the choice of supporting map ϕ for the carrying $\tau' < \tau$ even though the transformation of induced measures on τ' corresponding to measured geodesic laminations carried by τ' to induced measures on τ is independent of the choice of supporting map by Proposition 1.7.5. Observe, however, that incidence matrices are functorial in the sense that if $\tau'' < \tau'$ and $\tau' < \tau$ with respective incidence matrices M' and M for some choice of supporting maps, then there is a supporting map for the carrying $\tau'' < \tau$ whose incidence matrix is the product $M \, M'$. Incidence matrices are central to certain applications (see the Epilogue) and will be used at various points below.

Exercise: As we have observed, if τ' arises from τ by a split or shift, then $\tau' < \tau$. Compute the incidence matrices for the natural supporting maps of $\tau' < \tau$ in each case.

§2.2 EQUIVALENCE OF BIRECURRENT TRAIN TRACKS

In this section, we study the equivalence relation on measured train tracks introduced in the previous section in the setting of birecurrent train tracks, and all train tracks considered will be assumed to be birecurrent. To begin, if $\tau \subset F$ is birecurrent, then Construction 1.7.7 identifies the cone $V(\tau)$ with the collection of all measured geodesic laminations of compact support in F which are carried by τ by Theorem 1.7.12; we shall often make this identification here without further comment. Similarly, the polyhedron $U(\tau)$ is identified with the collection of all projectively measured geodesic laminations of compact support in F which are carried by τ, and $V(\tau)$ is identified with the cone on $U(\tau)$ from the empty lamination.

Of course, our definitions of splitting, shifting, and collapsing in the previous section apply to a positively measured birecurrent train track (τ, μ). As we have noted, if τ' arises from τ by a shift or a split, then $\tau' < \tau$, so transverse recurrence of τ implies transverse recurrence of τ' by Lemma 1.3.3b. We noted before that recurrence is invariant under refinement and therefore conclude

Proposition 2.2.1: *Birecurrence is invariant under refinement.*

On the other hand, collapsing can destroy transverse recurrence. For instance, the train track in Figure 1.3.2a is not transversely recurrent (as we have seen), while a split along the branch labeled b (for any choice of positive transverse measure on the train track) is necessarily a collision and produces a transversely recurrent train track. We shall show in Corollary 2.7.3 that any positively measured train track splits to a measured track which is transversely recurrent. Furthermore, in Corollary 2.3.4, we shall see that if two positively measured birecurrent train tracks are equivalent (where the intermediary train tracks might not be transversely recurrent), then they are equivalent through birecurrent train tracks.

124

Proposition 2.2.2 *Suppose that $\tau \subset F$ is a birecurrent train track.*
(a) In case τ' arises from τ by a shift, we have

$$U(\tau) = U(\tau') \text{ and } V(\tau) = V(\tau').$$

(b) If e is a large branch of τ, and τ'_R, τ'_O, and τ'_L, respectively, arise from a right split, a collision, and a left split of τ along e, then we have

$$U(\tau) = U(\tau'_L) \cup U(\tau'_O) \cup U(\tau'_R) \text{ and } V(\tau) = V(\tau'_L) \cup V(\tau'_O) \cup V(\tau'_R)$$

while

$$U(\tau'_O) = U(\tau'_R) \cap U(\tau'_L) \text{ and } V(\tau'_O) = V(\tau'_R) \cap V(\tau'_L),$$

where we adopt the conventions that $U(\sigma) = \emptyset$ and $V(\sigma)$ consists of the empty lamination if σ is not recurrent.

Proof: If τ' arises from τ by a shift, then $\tau' < \tau < \tau'$ so $V(\tau') \subset V(\tau) \subset V(\tau')$ by transitivity of carrying. Thus, $V(\tau) = V(\tau')$, proving part (a). As to part (b), we similarly conclude that $V(\tau'_L), V(\tau'_O), V(\tau'_R) \subset V(\tau)$. Furthermore, the three possibilities for splitting a measure on τ exhaust the possibilities for measures on τ, proving the first assertions of part (b). The second assertions of part (b) follow from the definition of splitting, and the proof is complete.   q.e.d.

The following result shows that our choice of equivalence relation is a natural one.

Proposition 2.2.3: *If (τ_1, μ_1) and (τ_2, μ_2) are equivalent birecurrent measured train tracks, then the measured geodesic laminations associated to (τ_1, μ_1) and (τ_2, μ_2) by Construction 1.7.7 coincide.*

Remark: The converse is also true (see Theorem 2.8.5), and the hypothesis of transverse recurrence can be dropped (see Theorem 2.7.4).

Proof: First observe that altering a measured train track by isotopy cannot change the associated measured geodesic lamination: indeed, an ambient isotopy lifts (and extends) to a map $\mathbb{H}^2 \cup S^1_\infty \rightarrow \mathbb{H}^2 \cup S^1_\infty$ which fixes S^1_∞ pointwise, so the underlying geodesic lamination is unchanged; the transverse measure produced in Step 3 of Construction 1.7.7 is furthermore defined in such a way that it is invariant as well. Suppose, then, that (τ_1, μ_1) arises from (τ_2, μ_2) by a single shift or split, where $\mu_1 > 0$. Consider the bi-foliated neighborhoods $(N_1, \mathcal{F}_1, \mathcal{F}_1^\perp)$ and $(N_2, \mathcal{F}_2, \mathcal{F}_2^\perp)$ of τ_1 and τ_2, respectively, which were defined in Step 1 of Construction 1.7.7. N_1 and N_2

agree except in a neighborhood of the branches involved in the splitting or shifting, and inspection of Figure 2.1.2 shows that μ_1 is derived from μ_2 in such a way that the corresponding measured laminations (defined in Step 2 of Construction 1.7.7 from the non-singular leaves of $\mathcal{F}_i$) are themselves isotopic. Thus, upon straightening to geodesic laminations (for which we require transverse recurrence), we arrive at the same measured geodesic lamination in each case, as was claimed. q.e.d.

To close this section, we investigate the relationship between $V(\tau')$ and $V(\tau)$ in case $\tau' < \tau$. Of course, by transitivity of carrying, $V(\tau') \subset V(\tau)$, and we have

Proposition 2.2.4: *Suppose that $\tau' < \tau \subset F$ are transversely recurrent train tracks, and let M denote the incidence matrix (cf. §2.1) corresponding to some choice of supporting map for the carrying of τ' by τ. The inclusion $V(\tau') \subset V(\tau)$ is given by the incidence matrix M in the sense that if $\mu' \in V(\tau')$, then Construction 1.7.7 associates the same measured geodesic lamination to (τ', μ') and $(\tau, M\mu')$ (where we regard a measure on τ or τ' as a tuple of weights on branches, as before).*

Proof: Let (G, λ) be the measured geodesic lamination and $(N', \mathcal{F}', \mathcal{F}^{\perp'})$ the bifoliated neighborhood associated to (τ', μ') by Construction 1.7.7. Furthermore, let $(N, \mathcal{F}, \mathcal{F}^{\perp})$ be the bifoliated neighborhood similarly associated to $(\tau, M\mu')$. Since $\tau' < \tau$, we may isotope N' inside N so that the ties of $\mathcal{F}^{\perp'}$ are restrictions of ties of $\mathcal{F}^{\perp}$, and since $G < \tau'$, we may isotope G so that it lies inside N' transverse to ties of $\mathcal{F}^{\perp'}$. Insofar as $G < \tau$, the measure λ on G gives rise to a measure μ on τ (as described after Corollary 1.7.4), and, by construction $\mu = M\mu'$. By Theorem 1.7.12, (G, λ) is associated to $(\tau, \mu) = (\tau, M\mu')$, and the proof is complete. q.e.d.

Remark: In contrast to this result, we recall from §2.1 that the incidence matrix itself depends on the choice of supporting map.

§2.3 SPLITTING VERSUS SHIFTING

In this section, we shall show that the equivalence relation generated by splitting (and isotopy) alone coincides with the equivalence relation (considered in the previous sections) generated by splitting and shifting (and isotopy). More precisely, we will show the following.

Theorem 2.3.1: *If (τ_1, μ_1) and (τ_2, μ_2) are equivalent positively measured train tracks, then there is a measured train track (τ, μ) so that (τ_i, μ_i) splits (and isotopes) to (τ, μ), for $i = 1, 2$.*

Before proving this result, we must discuss some generalities.

Suppose that $\rho \colon [0, N] \to \tau$ is a finite trainpath on the (recurrent and generic) track τ. We say that ρ is a *large trainpath* if the half-branch $\rho((0, \frac{1}{2}))$ is large at the trivalent switch $\rho(0)$ and the half-branch $\rho((N - \frac{1}{2}, N))$ is large at the trivalent switch $\rho(N)$. We say that ρ is a *one-way trainpath* provided that ρ is a large trainpath and $\rho((j - \frac{1}{2}, j))$ is small at $\rho(j)$, for each $j = 1, \ldots, N - 1$. See Figure 2.3.1a for an example of a one-way trainpath which begins with the branch labeled b and ends with the branch labeled b''.

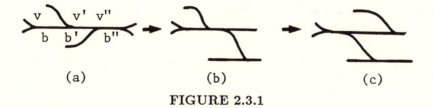

(a)   (b)   (c)

FIGURE 2.3.1

Lemma 2.3.2: *For each trivalent switch v of the recurrent generic track τ, there is a unique one-way trainpath on τ starting at v. This trainpath is necessarily imbedded.*

Proof: Uniqueness of one-way trainpaths is clear, and we first concentrate

127

on proving existence. There is a unique branch b of τ which is large at v.
If b is actually a large branch, then existence is trivial, and if b is only a
half-large branch, then let v' denote the switch of τ at which b is small. By
recurrence of τ, $v \neq v'$, and there is a unique branch b' of τ which is large
at v'. If b' is a large branch, then existence again follows easily, while if
b' is small at the switch v'', then $v'' \notin \{v, v'\}$ by recurrence, and there is
a unique branch $b'' \notin \{b, b'\}$ which is large at v''. Continuing in this way,
we produce ever-larger imbedded trainpaths whose first branch is large at
v. Since τ has only finitely many branches, this process terminates with an
imbedded one-way trainpath starting from v, proving the lemma.   q.e.d.

Suppose that (τ, μ) is a measured generic track with $\mu > 0$ and v is a
trivalent switch of τ. We next define a sequence

$$(\tau, \mu) = (\tau_0, \mu_0), (\tau_1, \mu_1), (\tau_2, \mu_2), \ldots$$

of measured tracks where (τ_{j+1}, μ_{j+1}) arises from (τ_j, μ_j) by a single split,
for each j. To define this sequence, begin with $j = 0$, and consider the one-
way trainpath ρ_j on τ_j starting at v. If this trainpath has length one, so that
there is a large branch incident on v, then the sequence of measured tracks
terminates with (τ_j, μ_j); in the contrary case, perform a split on the last
branch traversed by ρ_j (which is a large branch) to produce (τ_{j+1}, μ_{j+1}).
Since ρ_j is imbedded in τ_j by the previous lemma, τ_{j+1} agrees with τ_j in
a neighborhood of v, and we let v also denote the corresponding switch of
τ_{j+1}. Continuing in this way, we produce the desired finite or semi-infinite
sequence of measured tracks. We call this the *process of the trivalent switch
v of τ with respect to the measure μ*. An example of such a process is given
in Figure 2.3.1. In fact, this sequence of measured tracks is always finite,
as we next show.

Proposition 2.3.3: *If v is a trivalent switch of the generic track τ, then
the process of v with respect to the measure $\mu > 0$ on τ terminates.*

Proof: Since there can be at most a finite number of collisions during the
process of a trivalent switch, we may assume without loss that there are, in
fact, no collisions. Thus, each singular leaf in the interior of the bi-foliated
neighborhood $(N, \mathcal{F}, \mathcal{F}^\perp)$ associated to (τ, μ) in Step 1 of Construction 1.7.7
is semi-infinite and is equipped with a canonical orientation.

We associate a complexity to the triple (τ, μ, v) as follows. Let $q \geq 1$
denote the length of the one-way trainpath on τ which begins at v, and
define $p \geq q$ to be the infimum of the lengths of the tie-transverse paths of
$(N, \mathcal{F}, \mathcal{F}^\perp)$ corresponding to initial segments of singular leaves of $\mathcal{F}$ whose
associated trainpaths terminate by traversing the half-branch of τ which is

large at v. This infimum must be finite since otherwise N would contain an imbedded region of infinite area (for the natural metric on N where each rectangle comprising N has length unity and width given by the measure μ). Finally, the complexity of (τ, μ, v) is the ordered pair (p, q) with the lexicographic order.

Now, consider the process of v with respect to μ. There are two possibilities for a split from (τ_j, μ_j) to (τ_{j+1}, μ_{j+1}) (since we have assumed that there are no collisions) as in Figure 2.3.2a and b, which represent a right and left split, respectively. In the former case, p is either unchanged or decreased by one and q is decreased by one, while in the latter case, q may increase (subject to $p \geq q$, of course) but p must decrease by one. It follows that each split decreases the complexity, so the process of v must terminate, as was claimed. q.e.d.

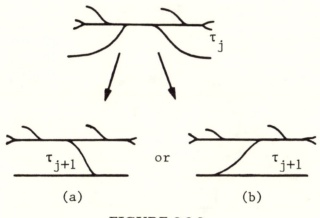

FIGURE 2.3.2

Remark: In effect, the previous proposition says that given a half-large branch b which is large at a switch v of a track τ and given a measure $\mu > 0$ on τ, we may split (τ, μ) outside a neighborhood of v until the branch of the resulting track corresponding to b is large.

Proof of Theorem 2.3.1 : To begin, we claim that if (τ_1, μ_1) and (τ_2, μ_2) are related by splitting and collapsing alone, then they split to a common measured track. Of course, if (τ_i, μ_i) arises from (τ_j, μ_j) by splitting alone, for $\{i, j\} = \{1, 2\}$, then the claim is trivial; similarly, if (τ_i, μ_i) splits to a measured track (τ', μ') which itself collapses to (τ_j, μ_j), for $\{i, j\} = \{1, 2\}$, then the claim follows by taking $(\tau, \mu) = (\tau', \mu')$. The salient point for completing the proof of the claim is that collapses and splits commute in the following sense: if (τ', μ') arises from (τ_i, μ_i) by a single collapse, for $i = 1, 2$, then either (τ_1, μ_1) is isotopic to (τ_2, μ_2), or there is some measured track (τ'', μ'') which collapses to both of (τ_i, μ_i), for $i = 1, 2$. Indeed, let b_i denote the branch of (τ', μ') along which one splits to produce (τ_i, μ_i), for $i = 1, 2$; if $b_1 = b_2$, then (τ_1, μ_1) is isotopic to (τ_2, μ_2), and if $b_1 \neq b_2$, then let (τ'', μ'') be the track obtained from (τ', μ') by performing splits along *both* of b_1 and b_2. It follows that the composition of a collapse followed by a split can be replaced by the composition of a split followed by a collapse, so any composition of collapses and splits can be put into one of the forms discussed above, proving the claim.

Next, consider the case in which (τ_1, μ_1) differs from (τ_2, μ_2) by a single shift and adopt the notation of Figure 2.3.3a, where the shift on τ_i is performed along a branch which is small at the switch v_i and large at the switch v_i', for $i = 1, 2$. Let (p_i, q_i) denote the complexity as defined in the proof of Proposition 2.3.3 above corresponding to the triple (τ_i, μ_i, v_i'), for $i = 1, 2$. Define a new lexicographically ordered "joint" complexity by $(p, q) = (p_1 + p_2, q_1 + q_2)$, and notice that $q_1 = q_2$. We prove the theorem in this case by induction on the complexity (p, q).

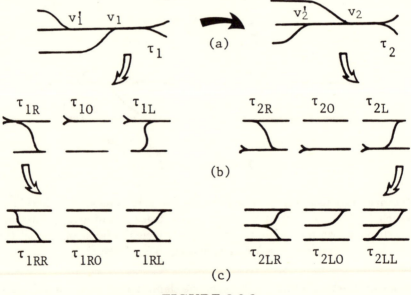

FIGURE 2.3.3

If $\frac{q}{2} = q_1 = q_2 > 2$, then split along the last branch of the one-way trainpath on τ_i starting at v_i', for $i = 1, 2$. The resulting tracks differ from one another by a single shift, and, as before, each of the complexities (p_1, q_1) and (p_2, q_2) are decreased. Thus, the joint complexity is decreased as well, and the induction step is valid.

For the basis step, we assume that $\frac{q}{2} = 2$ (so Figure 2.3.3a is accurate) and argue as follows. Split along the large branches in Figure 2.3.3a in all possible ways, and let τ_{iX} denote the resulting tracks as in Figure 2.3.3b, where $i = 1, 2$ and $X \in \{R, O, L\}$. Each of the splits to τ_{1R} and τ_{2L} introduces one new large branch, and we furthermore split along each of these branches in all possible ways to produce the tracks τ_{1RX} and τ_{2LX}, for $X \in \{R, O, L\}$, indicated in Figure 2.3.3c. Of course, the measures μ_1 and μ_2 on the original tracks uniquely determine which of these possible splits actually occurs, and there are induced measures on the resulting tracks. One easily checks that

τ_{1O} occurs if and only if τ_{2LO} occurs,
τ_{2O} occurs if and only if τ_{1RO} occurs,
τ_{1RL} occurs if and only if τ_{2LR} occurs,
τ_{1L} occurs if and only if τ_{2LL} occurs,

and

τ_{2R} occurs if and only if τ_{1RR} occurs.

Furthermore, one observes that

τ_{1O} is isotopic to τ_{2LO},
τ_{2O} is isotopic to τ_{1RO},

and

τ_{1RL} is isotopic to τ_{2LR},

while

τ_{1L} differs from τ_{2LL} by a single shift, and, in this case, $p_1 > 2$,

and

τ_{2R} differs from τ_{1RR} by a single shift, and, in this case, $p_2 > 2$.

Thus, unless the measures μ_1 and μ_2 are so that τ_{1L} or τ_{2R} occurs, then (τ_1, μ_1) and (τ_2, μ_2) indeed split to a common measured track, and the basis step would be complete. Furthermore, observe that the joint complexity associated to the shift of τ_{1L} to τ_{2LL} (or of τ_{2R} to τ_{1RR}) strictly precedes the joint complexity associated to the shift of τ_1 to τ_2 in the lexicographic order. (Indeed, each separate complexity is actually decreased.) It follows that these possibilities are untenable after sufficient splitting (since $p_1 > 2$ and $p_2 > 2$, respectively, in these cases), and the claim is proved.

To finally prove the theorem, suppose simply that (τ_1, μ_1) is equivalent to (τ_2, μ_2), and imagine applying a composition of splits, collapses, and shifts to (τ_1, μ_1) to produce (τ_2, μ_2). By the previous argument, any shift in the composition can be replaced by a composition of splits and collapses, so we may assume that (τ_1, μ_1) is related to (τ_2, μ_2) by splitting and collapsing alone. The argument which began the proof of this theorem then applies to show that (τ_1, μ_1) and (τ_2, μ_2) split to a common track, as was claimed.

q.e.d.

Corollary 2.3.4: *If (τ_1, μ_1) and (τ_2, μ_2) are birecurrent tracks which are equivalent through measured (not necessarily transversely recurrent) tracks, then (τ_1, μ_1) (τ_2, μ_2) are equivalent through birecurrent tracks.*

Proof: This follows from Theorem 2.3.3 since splitting preserves transverse recurrence.

q.e.d.

§2.4 EQUIVALENCE VERSUS CARRYING

We have already observed that if the track τ splits or shifts to the track τ', then $\tau > \tau'$. Since the relation $<$ is transitive, it follows that if the positively measured track (τ, μ) refines to the positively measured track (τ', μ'), then $\tau > \tau'$. Most of this section is dedicated to proving a useful converse of this result.

Suppose that τ is a train track carrying the measured geodesic lamination G, and let μ denote the corresponding measure on τ. Suppose that $\mu > 0$ and recall the bifoliated neighborhood $(N, \mathcal{F}, \mathcal{F}^{\perp})$ associated to (τ, μ) in Step 1 of Construction 1.7.7. We say that τ is *suited to* G provided that there is no finite singular leaf of $\mathcal{F}$ which lies in the interior of N. Under these circumstances, no collisions can occur in any refinement of (τ, μ).

We can now state the main result of this section as

Theorem 2.4.1: *Suppose that $\tau' < \tau$ are train tracks in F, τ is transversely recurrent, and the measure $\mu' > 0$ induces the measure $\mu > 0$ on τ. Let G be the measured geodesic lamination in F corresponding to (τ, μ) and suppose that $G < \tau'$. Then (τ, μ) refines to (τ', μ'). In particular, if τ is suited to G, then no collisions occur during this refinement.*

The proof will require the following general construction.

Suppose that $\mu > 0$ is a measure on the track τ, and let $(N, \mathcal{F}, \mathcal{F}^{\perp})$ be the corresponding bifoliated neighborhood. Let ς be a singular leaf of $\mathcal{F}$ which lies in the interior of N, and suppose first that ς is semi-infinite and issues forth from the singular point s. As in §1.7, we may parametrize $\varsigma \cup \{s\}$ by some injection

$$\Sigma \colon [0, \infty) \to N$$

so that $\Sigma(0) = s$ and $\Sigma(t)$ lies in a singular tie of $\mathcal{F}^{\perp}$ if and only if $t \in \mathbb{Z}_{+} \cup \{0\}$. For each $t \in [0, \infty)$, define

$$\varsigma(t) = \Sigma([0, t)), \text{ for } t \geq 0.$$

As before, the foliations $\mathcal{F}$ and $\mathcal{F}^{\perp}$ restrict to foliations $\mathcal{F}_t$ and $\mathcal{F}_t^{\perp}$, respec-

tively, of
$$N(t) = N - \varsigma(t),$$

and collapsing ties of $\mathcal{F}_t^\perp$ gives rise to a birecurrent train track $\tau(t) < \tau$; let $\mu(t) > 0$ denote the corresponding measure on $\tau(t)$. Provided $t \notin \mathbb{Z}_+$, the track $\tau(t)$ is generic, and if $t, t' \in [0, \infty) - \mathbb{Z}_+$ and there is no natural number between t and t', then the tracks $\tau(t)$ and $\tau(t')$ are isotopic. In particular, it follows that the isotopy class of $(\tau(t), \mu(t))$ is independent of the choice Σ of parametrization.

In case ς happens to be a finite singular leaf of $\mathcal{F}$ interior to N, then we may choose a singular point s in the closure $\bar{\varsigma}$ of ς, choose a parametrization

$$\Sigma \colon [0, M] \to N$$

of $\bar{\varsigma}$, for some $M \in \mathbb{Z}_+$, so that $\Sigma(0) = s$ and $\Sigma(t)$ lies in a singular tie of $\mathcal{F}^\perp$ if and only if $t \in \{0, 1, \ldots, M\}$. As before, we define

$$\varsigma(t) = \Sigma([0, t)), \text{ for } 0 \leq t \leq M$$

and extend this definition by taking

$$\varsigma(t) = \Sigma([0, M]), \text{ for } t > M.$$

The foliations on N again restrict to foliations of

$$N(t) = N - \varsigma(t),$$

and there is an associated measured track $(\tau(t), \mu(t))$. (Notice that we may have to add a bivalent vertex to simple closed curve components of $\tau(t)$.)

In any case, we say that the measured track $(\tau(t), \mu(t))$ arises from the measured track (τ, μ) by a *t-unzipping along ς (starting from s)*; we say simply that (τ', μ') arises by *unzipping (τ, μ) along ς* if it arises from a t-unzipping along ς, for some t. If ς is a finite singular leaf parametrized as above by a map $\Sigma : [0, M] \to N$, then the track $(\tau(t), \mu(t))$ for $t > M$ is said to arise from a *complete unzipping* of (τ, μ) along ς. Finally, we say simply that (τ, μ) *unzips* to (τ', μ') if there is a sequence $(\tau, \mu) = (\tau_1, \mu_1), (\tau_2, \mu_2), \ldots, (\tau_K, \mu_K) = (\tau', \mu')$ so that (τ_k, μ_k) unzips along some singular leaf of the corresponding bifoliated neighborhood to (τ_{k+1}, μ_{k+1}) for each $k = 1, \ldots, K - 1$. Of course, the relation "unzips to" is a transitive relation on measured tracks.

Lemma 2.4.2: *Suppose that τ is a generic birecurrent train track supporting a measure $\mu > 0$ and (τ, μ) unzips to (τ', μ'). Then $\mu' > 0$, and, if τ' is generic, then (τ, μ) refines to (τ', μ').*

Proof: To begin, notice that if (τ, μ) unzips along some singular leaf to the measured train track $(\tau(t_2), \mu(t_2))$ and $0 \leq t_1 \leq t_2$, then (τ, μ) unzips to $(\tau(t_1), \mu(t_1))$, and $(\tau(t_1), \mu(t_1))$ unzips to $(\tau(t_2), \mu(t_2))$. By transitivity of the relation "refines to", it therefore suffices to prove the lemma in case $(\tau', \mu') = (\tau(t), \mu(t))$ arises from the t-unzipping along some singular leaf ς for $t \in [0, \frac{3}{2}] - \{1\}$. In light of remarks above, this amounts to showing simply that (τ, μ) refines to $(\tau(\frac{3}{2}), \mu(\frac{3}{2}))$, and there are two cases depending on whether the first branch b traversed by the trainpath corresponding to ς is half-large or large: $(\tau(\frac{3}{2}), \mu(\frac{3}{2}))$ arises from (τ, μ) in the former case by a shift along b and, in the latter case, by a split along b.

That positivity of the measure μ implies positivity of the measure μ' is obvious, and the proof is complete.   q.e.d.

Proof of Theorem 2.4.1: Let $(N, \mathcal{F}, \mathcal{F}^{\perp})$ denote the bifoliated neighborhood associated to (τ, μ). Since $\tau' < \tau$, we may isotope τ' in F so that it lies in the interior $\overset{\circ}{N}$ of N and is transverse to the ties of $\mathcal{F}^{\perp}$. Furthermore, we may choose a regular neighborhood N' of τ' in $\overset{\circ}{N}$ so that $\mathcal{F}^{\perp}$ restricts to a foliation $\mathcal{F}'^{\perp}$ of N' and the quotient of N' by collapsing each leaf of $\mathcal{F}'^{\perp}$ to a point produces the track τ'; see Figure 2.4.1. Since G arises from (τ', μ'), $G < \tau'$ by Theorem 1.7.8, so we may isotope G to a measured lamination G' in F so that $G' \subset N'$ and G' is transverse to the ties of $\mathcal{F}^{\perp}$. The collapse of ties of $\mathcal{F}^{\perp}$ thus determines a supporting map for the carrying $G' < \tau$, and the induced measure on τ must agree with the measure μ on τ by unicity of measure in Theorem 1.7.12. In particular, since $\mu > 0$ by hypothesis, it follows that τ' maps onto τ under the collapse of ties of $\mathcal{F}^{\perp}$.

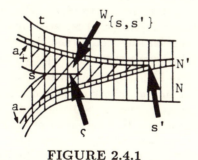

FIGURE 2.4.1

Now, fix a cusp s in the frontier of N, let t denote the corresponding singular tie of $\mathcal{F}^{\perp}$, and let ς denote the singular leaf of $\mathcal{F}$ issuing forth from s. Since τ' maps onto τ, there must be points of the frontier of N' on either side of s in t, and we consider the smooth arcs a_+ and a_- in the frontier of N' which meet t nearest s on either side of s. It may happen that as we traverse a_+ and a_- (away from s on the same side of t as ς), these arcs coalesce at a cusp s' of N', as in Figure 2.4.1. If not, then a_+ and a_- must traverse distinct paths and so there is some first cusp $\hat{s}$ of N so that, say, a_+ passes to the left of $\hat{s}$ and a_- passes to the right of $\hat{s}$ as in Figure 2.4.2a.

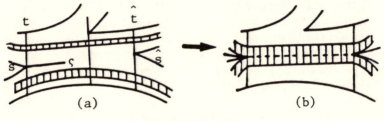

(a)   (b)

FIGURE 2.4.2

We claim that in the latter case, ς must be a finite singular leaf which is incident on the cusps s and $\hat{s}$; in particular, if τ is suited to G, then this case cannot occur. To see this, suppose that $\hat{s}$ lies on the singular tie $\hat{t}$ of $\mathcal{F}^\perp$, and take the quotient $\hat{N}$ of N' gotten by collapsing to a point each sub-arc of a tie connecting a_+ to a_- and lying between t and $\hat{t}$, as in Figure 2.4.2b. The foliation $\mathcal{F}'^\perp$ of N' gives rise to a foliation of $\hat{N}$, and the collapse of leaves of the foliation of $\hat{N}$ produces a train track $\hat{\tau}$, so that $\tau' < \hat{\tau} < \tau$. Of course, the measure μ' gives rise to a measure $\hat{\mu}$ on $\hat{\tau}$, and $\hat{\mu}$ in turn gives rise to the measure μ by unicity of measure. It follows that ς is indeed a finite singular leaf connecting s to $\hat{s}$, and, in fact, $(\hat{\tau}, \hat{\mu})$ arises from (τ, μ) by a complete unzipping along ς.

By Lemma 2.4.2, (τ, μ) refines to $(\hat{\tau}, \hat{\mu})$. Furthermore, the hypotheses of Theorem 2.4.1 hold with (τ, μ) replaced by $(\hat{\tau}, \hat{\mu})$. Since there are only finitely many singular leaves of $\mathcal{F}$ (and the relation of "refines to" is transitive), we may therefore assume by completely unzipping along all the finite singular leaves in $\overset{\circ}{N}$ that G is suited to τ, so for any singular point s of N, the paths a_+ and a_- above do indeed coalesce at some singular point s' of N'. Conversely, since $\mu' > 0$, one sees easily that each cusp s' of τ' is associated in this way to some cusp s of N. There is thus a pairing of the cusps of N with those of N', and each pair $\{s, s'\}$ in this pairing determines a region $W_{\{s,s'\}} \subset N - N'$ bounded by sub-arcs of a_+, a_-, and t; see Figure 2.4.1.

Consider the quotient N'' of N' gotten by collapsing to a point each sub-arc of a tie which is contained in some region $W_{\{s,s'\}}$. The foliation $\mathcal{F}'^\perp$ gives rise to a foliation $\mathcal{F}''^\perp$ of N'', and collapsing ties of $\mathcal{F}''^\perp$ gives rise to a track τ'' so that $\tau' < \tau'' < \tau$. Furthermore, the measure μ' induces a measure μ'' on τ'', and (τ'', μ'') unzips to (τ', μ') by construction. We claim that the collapse of leaves of $\mathcal{F}^\perp$ restricts to a homeomorphism of τ'' to τ, and it suffices to show that exactly one branch of τ'' intersects each tie of $\mathcal{F}^\perp$; in the contrary case, we can find a pair of adjacent such intersections, the corresponding arcs in the frontier of N'' diverge by construction, and we find an arc in $N - \tau''$ transverse to the ties connecting cusps of N, which is impossible, as before.

By unicity of measure, the measure μ'' on τ'' gives rise to the measure μ on τ, and we conclude that (τ, μ) unzips to (τ', μ'). Applying Lemma 2.4.2, we conclude that (τ, μ) refines to (τ', μ'), as was asserted. Finally, if τ is suited to G, then the pairing above establishes a one-to-one correspondence between the set of switches of τ and the set of switches of τ', so this refinement involves no collisions.   q.e.d.

Remark: This proof is borrowed from [PP1] though this result first appeared here (with a different proof).

Corollary 2.4.3: *Let* (τ, μ) *and* (τ', μ') *be two measured birecurrent generic train tracks which determine the same measured geodesic lamination. If* $\mu > 0$, $\mu' > 0$, *and* $\tau' < \tau$, *then* (τ, μ) *and* (τ', μ') *split to a common measured track.*

Proof: This follows immediately from Theorem 2.4.1 and Theorem 2.3.1.
                                         q.e.d.

Corollary 2.4.4: *Suppose that* (τ, μ) *and* (τ', μ') *are positively measured birecurrent generic train tracks. Then* (τ, μ) *refines to* (τ', μ') *if and only if* (τ, μ) *unzips to* (τ', μ').

Proof: If (τ, μ) refines to (τ', μ'), then the hypotheses of Theorem 2.4.1 are satisfied, and it follows from the proof that (τ, μ) unzips to (τ', μ'). The converse is simply a restatement of Lemma 2.4.2, and the proof is complete.
                                         q.e.d.

Remark: It follows easily from the previous result that refinement of unmeasured tracks (cf. §2.1) is transitive in the sense that if τ_1 refines to τ_2 and τ_2 refines to τ_3, then τ_1 refines to τ_3.

To conclude this section, we have the

Theorem 2.4.5 *Suppose that* (τ, μ) *is a positively measured birecurrent train track in* F *determining the measured geodesic lamination* G *in* F.

If G' is another (unmeasured) geodesic lamination in F which is not a sub-lamination of (the unmeasured geodesic lamination underlying) G, then (τ, μ) splits to some measured track (τ', μ') so that G' is not carried by τ'.

Proof: Let $(N, \mathcal{F}, \mathcal{F}^{\perp})$ denote the bifoliated neighborhood associated to (τ, μ), and apply Corollary 1.7.13 to construct a sequence $\tau = \tau_0 > \tau_1 > \dots$ of birecurrent generic train track so that $E(\tilde{G}) = \cap_{m \geq 0} E(\tau_m)$, where $\tilde{G}$ is the full pre-image of G in $\mathbb{H}^2$. Observe that each τ_m arises from τ by unzipping.

If G' is not a sub-lamination of G, then there is some leaf g' of G' which is not a leaf of G, so $E(\tilde{g}') \notin E(\tau_m)$ for some m, where $\tilde{g}'$ is any lift of g' to $\mathbb{H}^2$. Thus, $E(\tilde{G}') \not\subset E(\tau_m)$, where $\tilde{G}' \subset \mathbb{H}^2$ is the full-pre-image of G'. It follows that G' is not carried by τ_m by Theorem 1.6.6.

Finally, since (τ, μ) refines to (τ_m, μ_m) by the previous corollary, there is some track (τ', μ') so that each of (τ, μ) and (τ_m, μ_m) splits to (τ', μ') by Theorem 2.3.1. Since $\tau_m > \tau'$ and G' is not carried by τ_m, G' is likewise not carried by τ', as desired.   q.e.d.

§2.5 SPLITTING AND EFFICIENCY

We describe in this section an algorithm for splitting a measured track so as to hit any given multiple curve efficiently. We must begin with some generalities and suppose that $\tau \subset F$ is a train track and $C \subset F$ is a multiple curve meeting τ transversely (if at all). Suppose that $B \subset F$ is an immersed bigon whose frontier lies in $\tau \cup C$, and let $\beta \subset C$ denote the frontier edge of B in C. We say that the bigon B is *relevant* provided that the interior of β is imbedded in F; the frontier edge of B in τ may only be immersed. For example, the immersed bigons indicated in Figures 2.5.1a and 2.5.1b are relevant, while the one indicated in Figure 2.5.1c is not.

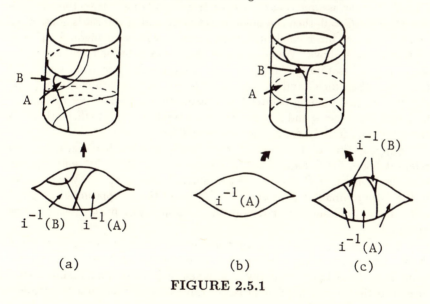

FIGURE 2.5.1

By Corollary 1.1.2, a relevant bigon is uniquely determined by its frontier edge β in C (whose endpoints lie in τ), and it follows that there can be only a finite number of relevant bigons for any fixed track τ and multiple curve C. Suppose that C does not hit τ efficiently, let $B \subset F$ be a bigon

139

with its frontier lying in $C \cup \tau$, as above, and consider the full pre-image $\tilde{B}$ of B in $\mathbb{H}^2$. If $\tilde{B}$ is an innermost such bigon, then it contains no branches of the full pre-image $\tilde{\tau}$ of τ in $\mathbb{H}^2$. Thus, B is imbedded in F as in Proposition 1.3.2, and we furthermore claim that B is relevant. To see this, let $\tilde{\beta}$ denote the frontier edge of $\tilde{B}$ lying in the lift $\tilde{C}$ of C to $\mathbb{H}^2$ and let $\tilde{p}$ be one of the vertices of $\tilde{B}$. If the covering projection is not an imbedding on the interior of $\tilde{\beta}$, then there must be another point $\tilde{q}$ in the interior of $\tilde{\beta}$ which projects to the same point as $\tilde{p}$. Thus, $\tilde{\tau}$ meets $\tilde{C}$ transversely at $\tilde{q}$, and this contradicts that $\tilde{B} \cap \tilde{\tau} = \emptyset$. Our claim is therefore proved, and we conclude that if C and τ do not hit efficiently, then there must be an imbedded relevant bigon for C and τ.

The algorithm mentioned above depends on a method of decreasing the number of relevant bigons and is described in the course of the proof of

Theorem 2.5.1 *If $C \subset F$ is a multiple curve and (τ, μ) is a positively measured generic train track in F, then (τ, μ) may be split to a measured track (τ', μ') and C may be isotoped to a multiple curve C' so that C' hits τ' efficiently.*

Proof: We may assume that C meets τ transversely away from the switches if at all. If $\tau \cap C = \emptyset$, then the theorem is trivial, and, if this set is non-empty and C does not hit τ efficiently, then there is a relevant bigon for C and τ, as discussed above. We shall prove the theorem by first refining (τ, μ) to (τ', μ') and isotoping C to C' so that there are fewer relevant bigons for C' and τ' than for C and τ. (We shall be concerned with splitting (τ, μ) rather than just refining it later.) Repeating this procedure, we ultimately find a multiple curve and a measured train track which hit efficiently since there are only finitely many relevant bigons for C and τ.

Suppose that C does not hit τ efficiently and let B be an imbedded relevant bigon for C and τ. Let $\tilde{B}$ denote a lift of B to $\mathbb{H}^2$ and suppose that the frontier of $\tilde{B}$ consists of two smooth arcs $\tilde{\alpha}$ and $\tilde{\beta}$, where $\tilde{\alpha}$ lies in the full pre-image $\tilde{\tau} \subset \mathbb{H}^2$ of τ and $\tilde{\beta}$ lies in the full pre-image $\tilde{C} \subset \mathbb{H}^2$ of C. Let $\alpha \subset C$ and $\beta \subset \tau$ denote the corresponding projections of these arcs to F.

We begin by considering the case in which the covering projection is an imbedding on the closure of $\tilde{\beta}$. Repeatedly splitting (equiviantly) along large branches of $\tilde{\tau}$ (if any) contained in $\tilde{\alpha}$ using the measure μ on τ decreases the (finite) number of large trainpaths in $\tilde{\tau}$ which are contained in $\tilde{\alpha}$. Thus, after sufficient splitting, we may assume that there are no large branches of $\tilde{\tau}$ contained in $\tilde{\alpha}$.

It may be that there are no branches of $\tilde{\tau}$ incident on $\tilde{\alpha}$; this occurs if $\tilde{\alpha}$ is contained in a branch of $\tilde{\tau}$ or perhaps contains a bivalent switch. In either case, the closure of α is imbedded in F. Thus, B is imbedded in F (since it is innermost), the closure of β is imbedded in F (by hypothesis), and the closure of α is imbedded in F (as above), so the closure of B is actually imbedded in F. One isotopes C across B (decreasing by two the number of intersections of the track and the multiple curve) as in Figure 2.5.2 to produce a multiple curve C' so that there are fewer relevant bigons for C' and $\tau' = \tau$ than for C and τ.

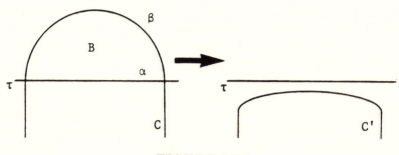

FIGURE 2.5.2

If there is a branch of $\tilde{\tau}$ incident on $\tilde{\alpha}$, then there are two possibilities for $\tilde{\tau}$ near $\tilde{B}$ as follows: either all the branches of $\tilde{\tau}$ in $\tilde{\alpha}$ are mixed (as in Figure 2.5.3a), or exactly one branch $\tilde{b}$ of $\tilde{\tau}$ in $\tilde{\alpha}$ is small and the remaining branches (if any) are mixed (as in Figure 2.5.3b).

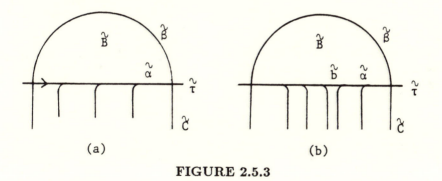

(a)   (b)

FIGURE 2.5.3

Concentrating first on the former case, we claim that the closure of $\tilde{\alpha}$ imbeds in F under the covering projection. To see this, let us orient $\tilde{\alpha}$ so that the induced orientation on each branch begins with a large half-branch and ends with a small half-branch (as in Figure 2.5.3a). Suppose that there are distinct points $\tilde{p}_i \in \tilde{\alpha}$ lying in branches $\tilde{b}_i$, for $i = 1, 2$, so that $\gamma(\tilde{p}_1) = \tilde{p}_2$ for some covering translation γ. The translation γ must satisfy

$\gamma(\tilde{b}_1) = \tilde{b}_2$ and preserve the orientations on $\tilde{b}_1$ and $\tilde{b}_2$ since it preserves the type (small or large) of half-branches. It follows that τ must contain a closed trainpath consisting entirely of mixed branches, and this contradicts our assumption that τ is recurrent (by Proposition 1.3.1 as τ supports a positive measure). On the other hand, suppose that there are distinct switches $\tilde{v}_i \in \tilde{\alpha}$ so that the branches $\tilde{b}_i$ of $\tilde{\tau}$ are large at $\tilde{v}_i$, for $i = 1, 2$, and a covering translation γ so that $\gamma(\tilde{v}_1) = \tilde{v}_2$. It follows that $\gamma(\tilde{b}_1) = \tilde{b}_2$, and this leads to a contradication, as above. Of course, a covering translation cannot identify a switch with a point in a branch of $\tilde{\tau}$, and we conclude that the closure of α is indeed imbedded in F.

We may therefore repeatedly shift the branches of τ incident on α using the measure μ to produce a measured track (τ', μ') so that there is exactly one branch of τ' incident on α; see Figure 2.5.4a. As before, we conclude that the closure of B is actually imbedded in F, isotope C across B (decreasing by one the number of intersections of the track and the multiple curve) as in Figure 2.5.4b to produce a multiple curve C' so that there are fewer relevant bigons for C' and τ' than for C and τ.

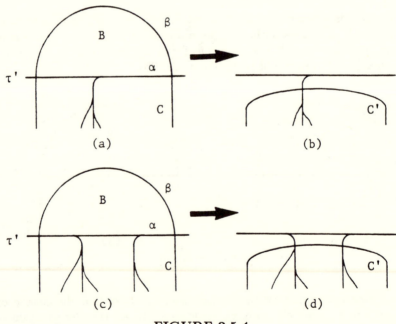

FIGURE 2.5.4

Turning to the case that $\tilde{\alpha}$ contains a small branch $\tilde{b}$, we may argue as above to conclude that the closure of each component of $\tilde{\alpha} - \tilde{b}$ imbeds in F under the covering projection. We may therefore repeatedly shift $\tilde{\tau}$ (equivariantly) using the measure μ to produce a train track $\tilde{\tau}'$ so that there are exactly two branches of $\tilde{\tau}'$ incident on $\tilde{\alpha}$ (cf. Figure 2.5.4c). We claim that the closure of α is imbedded in F. To see this, observe first that no covering translation can identify the two switches of $\tilde{\tau}'$ in $\tilde{\alpha}$, for otherwise, there would be a closed trainpath in the projection $\tau' \subset F$ of $\tilde{\tau}'$ of length one beginning and ending with a small half-branch, which is absurd. Thus, if the closure of α is not imbedded in F, then it projects to a closed imbedded trainpath of length two in τ' consisting of one small branch and one large branch, and this is easily seen to contradict that C is imbedded.

Arguing as before, we conclude that the closure of B is imbedded in F as in Figure 1.5.4c, and we isotope C across B to produce a multiple curve C' (meeting τ' in the same number of points as C) so that there are fewer relevant bigons for C' and τ' than for C and τ.

It remains to consider the case in which the closure of β is not imbedded in F, and it follows that the closure of β coincides with a component of C as in Figure 2.5.5a. Repeatedly splitting along large branches of τ in α (if any) using the measure μ, we can again arrange that α contains no large branches as in Figure 2.5.5b. Isotoping and shifting as in Figure 2.5.5c and 2.5.5d using the measure μ to produce a measured track (τ', μ'), we may furthermore arrange that there is at most one branch of τ' incident on α. Finally, isotope C across B (perhaps decreasing by one the number of intersections of the track and the multiple curve) as in Figure 2.5.5e to produce a multiple curve C' so that there are fewer relevant bigons for C' and τ' than for C and τ.

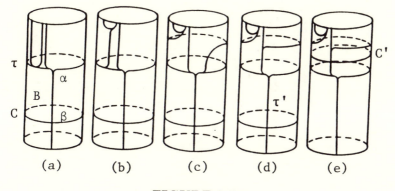

FIGURE 2.5.5

Thus, we may refine (τ, μ) to a measured track (τ', μ') and isotope

C to a multiple curve C' so that C' hits τ' efficiently, as was claimed. To complete the proof, we must show that (τ, μ) actually splits to a train track with this property. By Theorem 2.3.1, each of (τ, μ) and (τ', μ') split to a common measured train track (τ'', μ''), so it suffices to show that C' may be isotoped to a multiple curve C'' which hits τ'' efficiently. To this end, $\tau'' < \tau'$ since (τ', μ') splits to (τ'', μ''), and we may argue as in the proof of Lemma 1.3.3b to produce the required multiple curve C''.

Thus, C is isotopic to C'' and (τ, μ) splits to (τ'', μ'') so that C'' hits τ'' efficiently, and the proof is complete.   q.e.d.

Remark: There is a small subtlety in extending this result to the setting of train tracks with stops, as follows. Given a generic measured train track (τ, μ) with stops and a multiple arc C in F, we may have to perform inverse combs at the stops of τ to produce a non-generic measured track with stops which hits C efficiently. Such a situation is indicated in Figure 2.5.6.

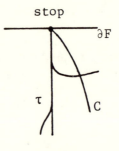

FIGURE 2.5.6

§2.6 THE STANDARD MODELS

The definition of the standard train tracks depends on a choice of basis for multiple curves as in Dehn's Theorem (see §1.2), and we choose the standard basis $\mathcal{A} = \mathcal{A}_g^s$ on $F = F_g^s$ depicted in Figure 1.2.5 once and for all. Let

$$\nu_i : A \to A_i, \quad i = 1,\dots,N = 3g - 3 + s$$
$$f_j : P \to P_j, \quad j = 1,\dots,M = 2g - 2 + s$$

be the characteristic maps for $\mathcal{A}$. Roughly, a train track $\tau \subset F$ is standard if each $\nu_i^{-1}(\tau) \subset A$ and $f_j^{-1}(\tau) \subset P$ coincides with certain canonical building blocks, which we next define.

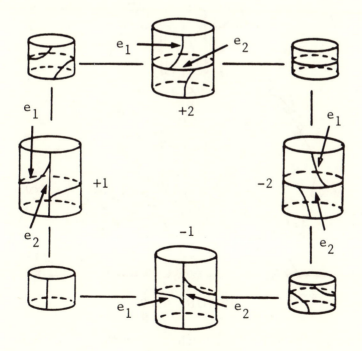

FIGURE 2.6.1

145

Begin by defining four maximal train tracks with stops (as in §1.8) in the standard annulus A, a surface of type $F_0^{0,2}$ in the notation of §1.8. These are depicted in Figure 2.6.1 and are called *(maximal) connectors of type* -2, -1, $+1$, or $+2$, as indicated; we also illustrate in Figure 2.6.1 the various recurrent subtracks of the maximal connectors; each such track is called simply a *connector*. Observe that each maximal connector has exactly two connectors as proper subtrack, and each such subtrack has exactly two maximal connectors as supertrack. In each maximal connector, we single out two branches e_1 and e_2 as indicated. Notice that the measures on e_1 and e_2 can be specified independently and uniquely determine a measure on the maximal connector (via the switch conditions) in each case. This labeling induces a labeling on certain branches of subtracks of the maximal connectors in the natural way, where, if $b \subset b'$ are branches of $\tau \supset \tau'$, respectively, then the label on b (if any) is associated also to b'.

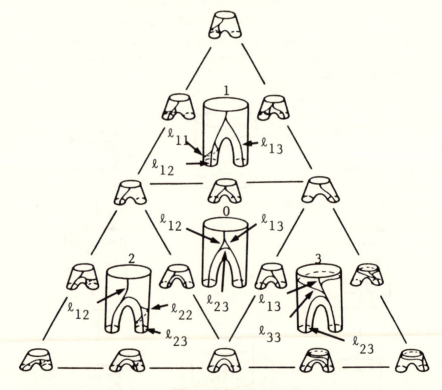

FIGURE 2.6.2

We also define four maximal train tracks with stops in the standard pants P. (Refer to Figure 1.2.2a to recall our notation for the various boundary components of P.) These are depicted in Figure 2.6.2 and are called *(maximal) basic tracks of type* 0, 1, 2, or 3 *in* P, as indicated. As before, we also illustrate each of the recurrent subtracks of the maximal basic tracks in P; each such track is called simply a *basic track in* P. For each maximal basic track in P, we single out three branches labeled by an appropriate subset of the collection of symbols

$$\{\ell_{ij} : 1 \leq i \leq j \leq 3\},$$

as indicated, and observe that the measures on these distinguished branches can be specified independently and uniquely determine a measure on the basic track in each case. As before, this labeling induces a labeling on certain branches of the (non-maximal) basic tracks.

A basic track τ in P is said to be of *archetype k* if τ carries a multiple arc in P which has both its endpoints in the boundary component ∂_k of P, for $k = 1, 2, 3$. Thus, there are exactly four basic tracks in P of archetype k, and exactly one of these is maximal (namely, the basic track of type k), for $k = 1, 2, 3$.

Now, consider a train track $\tau \subset F$ so that the following conditions are satisfied.

> (i) $f_j^{-1}(\tau)$ is either empty or is one of the basic tracks in the standard pants P, for each $j = 1, \ldots, M$.

> (ii) $\nu_i^{-1}(\tau)$ is either empty or is one of the connectors in the standard annulus A, for each $i = 1, \ldots, N$.

There are such train tracks τ which are not recurrent. For instance, if K is a pants curve in $\mathcal{A}$ which bounds a torus-minus-a-disk with corresponding characteristic map f mapping the component ∂_1 of ∂P to K, then any train track $\tau \subset F$ with $f^{-1}(\tau)$ of archetype 2 or 3 is necessarily not recurrent; see Figure 2.6.3a. In the same way, if K and K' are pants curves in $\mathcal{A}$ which bound a torus-minus-two-disks with corresponding characteristic maps f and f' so that $f(\partial_j)$ is isotopic to $f'(\partial_k)$, for $\{j, k\} = \{2, 3\}$, then any train track $\tau \subset F$ for which $f^{-1}(\tau)$ and $f'^{-1}(\tau)$ are each of archetype 2 or each of archetype 3 is necessarily not recurrent; see Figures 2.6.3b. We say a train track is *exceptional* if conditions (i)-(ii) above hold and there are pants curves in the basis $\mathcal{A}$ as described in this paragraph.

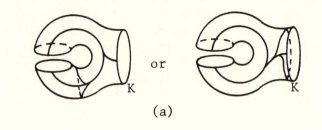

(a)

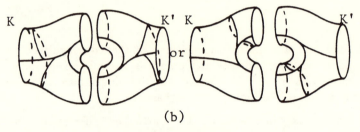

(b)

FIGURE 2.6.3

We treat separately the surface $F = F_1^1$, and say the train track $\tau \subset F$ is *standard* if conditions (i)-(ii) above hold, and, if $f : P \rightarrow P_1$ is the characteristic map of the pair of pants $P_1 \subset F$ (where we assume that the boundary component ∂_1 corresponds to the puncture of F), then $f^{-1}(\tau)$ is either empty or an arc connecting the components ∂_2 and ∂_3 in P.

In general, a train track $\tau \subset F \neq F_1^1$ satisfying conditions (i)-(ii) above is said to be *standard* if it is not exceptional.

A standard train track is *maximal* if it is not a proper subtrack of any standard track.

Lemma 2.6.1: *Any standard train track is birecurrent, and, in particular, any maximal standard train track is complete.*

Proof: In case $F = F_1^1$, the lemma is evidently true, and we assume henceforth that $F \neq F_1^1$.

By construction, a maximal standard train track in $F \neq F_1^1$ has only trigons and once-punctured monogons as complementary regions, so completeness of a maximal standard track follows from birecurrence.

We begin by showing that any standard train track is transversely recurrent, and, in light of Lemma 1.3.3a, it suffices to show that any maximal standard train track $\tau \subset F$ has this property. By Corollary 1.3.5, it suffices to exhibit a positive even tangential measure ν on τ, and, if b is a branch of τ, then we define

$$
\nu(b) = \begin{cases}
6, & \text{if } f_j^{-1}(\tau) \text{ is of archetype 1 and } f_j^{-1}(b) \text{ is one of } \ell_{13} \text{ or } \ell_{11}; \\
6, & \text{if } f_j^{-1}(\tau) \text{ is of archetype 2 and } f_j^{-1}(b) \text{ is one of } \ell_{12} \text{ or } \ell_{22}; \\
6, & \text{if } f_j^{-1}(\tau) \text{ is of archetype 3 and } f_j^{-1}(b) \text{ is one of } \ell_{23} \text{ or } \ell_{33}; \\
2, & \text{otherwise.}
\end{cases}
$$

We leave it as an elementary exercise to verify that ν is indeed a suitable tangential measure, completing the proof that any standard train track is transversely recurrent.

We turn now to recurrence of the standard tracks. Inspection of Figure 1.2.5 shows that the pants decomposition associated with our choice of basis for multiple curves in F gives rise to a corresponding decomposition of F into regions which are homeomorphic to $F_1^{0,1}$, $F_1^{0,2}$, $F_1^{1,1}$, $F_0^{1,2}$, or $F_0^{2,1}$, and any two such regions are either disjoint or meet along a single common pants curve. Using Dehn's Theorem for multiple arcs (as in §1.8), one easily concludes that the restriction of any standard track to one of these regions is a recurrent train track with stops. Finally, one combines multiple arcs on these regions to produce a multiple curve carried by a standard track τ in F with positive measure, and recurrence of τ finally follows from Proposition 1.3.1, completing the proof. q.e.d.

Recall from §2.1 that each cone $V(\tau)$ of measures on a recurrent train track τ has the natural structure of a cone on a finite-sided polyhedron, where faces of $V(\tau)$ are of the form $V(\sigma) \subset V(\tau)$ if $\sigma \subset \tau$ is a recurrent subtrack. We wish to study these faces when τ and hence σ are standard train tracks. Inspection of Figure 2.6.1 shows that the connectors in A may be chosen carefully to guarantee that any two intersect (as point sets) in a common subtrack, and the same applies to the basic tracks in P. We assume from now on that the connectors and basic tracks are chosen so that this is the case.

In order to study the faces of $V(\tau)$ for $\tau \subset F$ standard, we introduce new coordinates (also from [P1]) on $V(\tau)$. To begin, given a measure μ on the standard train track τ, define quantities ϵ_t^i for each $i = 1, \ldots, N$ and $t = 1, 2$ by

$$
\epsilon_2^i(\tau, \mu) = \begin{cases}
\mu(\nu_i(e_2)), & \text{if } e_2 \subset \nu_i^{-1}(\tau); \\
0, & \text{otherwise;}
\end{cases}
$$

$$
|\epsilon_1^i(\tau, \mu)| = \begin{cases}
\mu(\nu_i(e_1)), & \text{if } e_1 \subset \nu_i^{-1}(\tau); \\
0, & \text{otherwise,}
\end{cases}
$$

where the sign of ϵ_1^i is equal to the sign of the type of a maximal connector with $\nu_i^{-1}(\tau)$ as subtrack; notice that this sign is well-defined. We also define quantities λ_{mn}^j for $j = 1, \ldots, M$ and $1 \leq m \leq n \leq 3$ by

$$\lambda_{mn}^j(\tau, \mu) = \begin{cases} \mu(f_j(\ell_{m,n})), & \text{if } \ell_{mn} \subset f_j^{-1}(\tau); \\ 0, & \text{otherwise.} \end{cases}$$

For instance, if $f_j^{-1}(\tau)$ is of type 0 in P_j, then $\lambda_{kk}^j = 0$, for $k = 1, 2, 3$.

Now, define

$$Q_\tau : V(\tau) \to \mathrm{I\!R}^{6M+2N}$$
$$\mu \mapsto (\lambda_{11}^1, \lambda_{12}^1, \ldots, \lambda_{33}^1, \lambda_{11}^2, \ldots, \lambda_{33}^M, \epsilon_1^1, \epsilon_2^1, \epsilon_1^2, \ldots, \epsilon_2^N),$$

where we have suppressed the argument (τ, μ) of each of the coordinate functions of Q_τ. In light of previous remarks, the map Q_τ is a homeomorphism onto its image. The pleasant point about our new coordinates on $V(\tau)$ is that if $\sigma \subset \tau$ are standard train tracks, then $Q_\sigma(V(\sigma)) \subset Q_\tau(V(\tau))$ is determined by setting some collection of coordinates $\{\lambda_{mn}^j, \epsilon_t^i\}$ (which are non-zero on $V(\tau)$) equal to zero. Moreover, any such collection of equations determines a face of $V(\tau)$.

Recall that if $\mu \in V(\tau)$, then $\tau_\mu \subset \tau$ is the subtrack consisting of branches on which μ is non-vanishing, and the measure μ induces a positive measure on τ_μ. We define a complex $\mathcal{K}(F)$ by taking the quotient of the disjoint union of $\{V(\tau) : \tau \subset F \text{ is standard}\}$, where we identify (τ, μ) with (τ', μ') if $\tau_\mu = \tau'_{\mu'}$ (as point sets) and μ and μ' induce the same measure on $\tau_\mu = \tau'_{\mu'}$.

The collection $\{Q_\tau : \tau \subset F \text{ is standard}\}$ combine to give a map

$$q : \mathcal{K}(F) \to \mathrm{I\!R}^{6M+2N},$$

and we claim

Proposition 2.6.2 *The map q is a homeomorphism onto its image.*

Proof: To see that q is one-to-one, we describe the canonical construction of a standard measured train track (τ, μ) from parameters $\{\lambda_{mn}^j, \epsilon_t^i\}$ in the image of q. It is easy to see that one can uniquely construct $\tau \cap (\cup\{P_j\}_{j=1}^N)$ from $\{\lambda_{mn}^j\}$, and we concentrate on building the connector $\tau_i = \nu_i^{-1}(\tau)$ regarded as a train track with stops in A.

First, observe that the parameters $\{\lambda_{mn}^j\}$ themselves determine a quantity, the "intersection number with a component of $\bar{A}_i \cap \bar{P}_j$", according to the rule

$$m_{ij} = \begin{cases} 2\lambda_{11}^j + \lambda_{12}^j + \lambda_{13}^j, & \text{if } f_j(\partial_1) \text{ is homotopic to } K_i; \\ 2\lambda_{22}^j + \lambda_{12}^j + \lambda_{23}^j, & \text{if } f_j(\partial_2) \text{ is homotopic to } K_i; \\ 2\lambda_{33}^j + \lambda_{13}^j + \lambda_{23}^j, & \text{if } f_j(\partial_3) \text{ is homotopic to } K_i; \\ 0, & \text{otherwise.} \end{cases}$$

(Notice that if $\bar{A}_i \cap \bar{P}_j$ consists of two curves, then m_{ij} is still well-defined since the parameters lie in the image of q.) On the other hand, in order that the parameters $\{\lambda^j_{mn}, \epsilon^i_t\}$ arise from some standard train track, we must have

$$
m_{ij} = \begin{cases}
|\epsilon_1|, & \text{if } \tau_i \text{ is a maximal connector of type +2 or -2;} \\
|\epsilon_1| + \epsilon_2, & \text{if } \tau_i \text{ is a maximal connector of type +1 or -1;} \\
0, & \text{if } \tau_i \text{ is a subtrack of connectors of types -2 and +2;} \\
|\epsilon_1|, & \text{if } \tau_i \text{ is a subtrack of connectors of types -2 and -1;} \\
\epsilon_2, & \text{if } \tau_i \text{ is a subtrack of connectors of types +1 and -1;} \\
|\epsilon_1|, & \text{if } \tau_i \text{ is a subtrack of connectors of types +1 and +2,}
\end{cases}
$$

where $\epsilon_t = \epsilon^i_t$. Observe that the function m_{ij} extends continuously across faces of $V(\tau)$ in $\mathcal{K}(F)$.

From the formulas for m_{ij}, one sees that $m_{ij} \geq |\epsilon_1|$ in any case. Furthermore $m_{ij} > |\epsilon_1|$, if and only if τ_i is a (not necessarily proper) subtrack of a maximal connector of type +1 or -1. In this case, τ_i is maximal of type -1 if $\epsilon_1 < 0$, τ_i is maximal of type +1 if $\epsilon_1 > 0$, and τ_i is the common subtrack of these two maximal connectors if $\epsilon_1 = 0$.

If $m_{ij} = |\epsilon_1|$, then τ_i is a (not necessarily proper) subtrack of a maximal connector of type +2 or -2. If both of ϵ_1 and ϵ_2 are non-zero, then τ_i is of type twice the sign of ϵ_1. If $\epsilon_2 = 0$ and $\epsilon_1 \neq 0$, then τ_i is the common subtrack of type +1 and type +2 (if $\epsilon_1 > 0$) or the common subtrack of type -1 and type -2 (if $\epsilon_1 < 0$). Finally, if $\epsilon_1 = 0$, then τ_i is the common subtrack of type -2 and type +2.

In any case, we can uniquely reconstruct the measured standard track (τ, μ) from the parameters, and it follows that q is injective. As a consequence of the argument, we see that if τ is standard, then $Q_\tau(V(\tau)) \subset q(\mathcal{K}(F)) \subset \mathbb{R}^{6M+2N}$ is determined by requiring some collection of parameters among $\{\lambda^j_{mn}, \epsilon^i_t\}$ to vanish. It follows easily from this that q is continuous on $\mathcal{K}(F)$, and the proposition is then a consequence of invariance of domain.   q.e.d.

We refer to the parameters $\{\lambda^j_{mn}, \epsilon^i_t\}$ as *basic parameters* on $\mathcal{K}(F)$. The constraints on basic parameters which guarantee that they lie in the image of q is somewhat unpleasant, and the details of this are left to the reader. In practice, the unpleasantness of these conditions is usually avoided by considering other coordinates (from [P1]) on $\mathcal{K}(F)$ as follows. If K_i is a pants curve in the basis $\mathcal{A}$, $\bar{A}_i \cap \bar{P}_j \neq \emptyset$, (τ, μ) is a standard measured train track, and the hybrid intersection numbers m_{ij} and the train tracks $\tau_i \subset A$

are defined as above, then we define

$$m_i(\tau, \mu) = m_{ij}$$

$$\bar{t}_i(\tau, \mu) = \begin{cases} -(|\epsilon_1| + \epsilon_2), & \text{if } \tau_i \text{ is of type } -2; \\ -|\epsilon_1|, & \text{if } \tau_i \text{ is of type } -1; \\ +|\epsilon_1|, & \text{if } \tau_i \text{ is of type } +1; \\ +|\epsilon_1| + \epsilon_2, & \text{if } \tau_i \text{ is of type } +2, \end{cases}$$

where $\epsilon_t = \epsilon_t^i(\tau, \mu)$, for $t = 1, 2$. We have seen above that m_i is continuous on $\mathcal{K}(F)$ and remark that $\bar{t}_i$ extends to a function on $\mathcal{K}(F)$ which is continuous except on the locus where τ_i is the common subtrack of type $+2$ and type -2, that is, when $m_i = 0$ and $\epsilon_2 \neq 0$. To take account of this, we introduce

$$t_i(\tau, \mu) = \begin{cases} \bar{t}_i(\tau, \mu), & \text{if } m_i(\tau, \mu) \neq 0; \\ |\bar{t}_i(\tau, \mu)|, & \text{if } m_i(\tau, \mu) = 0; \end{cases}$$

and define

$$\Pi : \mathcal{K}(F) \to \mathcal{H} = \left[(\mathbb{R}_+ \times \mathbb{R}) \cup (\{0\} \times (\mathbb{R}_+ \cup \{0\})) \right]^N$$

$$(\tau, \mu) \mapsto (m_1(\tau, \mu), t_1(\tau, \mu), \ldots, m_N(\tau, \mu), t_N(\tau, \mu)).$$

The quantities m_i and t_i are called the ith "intersection number" and "twisting number", respectively; we shall refer to these as the *Dehn-Thurston coordinates* on $\mathcal{K}(F)$.

Finally, define π to be the composition

$$\mathcal{K}(F) \to \mathcal{H} \to \left(\mathbb{R}^2/\text{antipodal map}\right)^N$$

where the first map is Π, and the second is the natural quotient map.

Proposition 2.6.3 *The map* $\pi : \mathcal{K}(F) \to \mathbb{R}^{2N}$ *is a homeomorphism.*

Proof: It suffices to give a continuous inverse to the continuous map from the basic parameters to the Dehn-Thurston coordinates on $\mathcal{K}(F)$. We establish here the convention that the sign of zero is taken to be unity and define

$$\text{sign}(\epsilon_1^i) = \text{sign}(t_i)$$

$$|\epsilon_1^i| = \begin{cases} m_i, & \text{if } \tau_i \text{ is a subtrack of type } +2 \text{ or type } -2; \\ |t_i|, & \text{if } \tau_i \text{ is a subtrack of } +1 \text{ or type } -1; \end{cases}$$

$$\epsilon_2^i = ||t_i| - m_i|,$$

where, as before, $\tau_i = \nu_i^{-1}(\tau)$. Furthermore, if P_j is a pair of pants in the basis $\mathcal{A}$ whose boundary components correspond to (not necessarily distinct) pants curves in $\mathcal{A}$, say with intersection number m_k corresponding to $f_j(\partial_k)$, for $k = 1, 2, 3$, then

$$2\lambda_{11}^j = (m_1 - m_2 - m_3) \vee 0$$
$$2\lambda_{12}^j = (m_1 + m_2 - m_3) \vee 0$$
$$2\lambda_{13}^j = (m_1 + m_3 - m_2) \vee 0$$
$$2\lambda_{22}^j = (m_2 - m_1 - m_3) \vee 0$$
$$2\lambda_{23}^j = (m_2 + m_3 - m_1) \vee 0$$
$$2\lambda_{33}^j = (m_3 - m_1 - m_2) \vee 0,$$

where $\vee$ denotes the binary supremum.

We leave it as an exercise to check that the formulas given above correspond to a continuous inverse to the map π, as was claimed.   q.e.d.

§2.7 EXISTENCE OF THE STANDARD MODELS

Our next task is to show that any measured train track (without stops) in $F = F_g^s$ is equivalent to a standard one. (In the next section, we will show that this standard train track is unique.) In preparation for this, we must first consider train tracks with stops in a surface F of type $F_0^{0,3}$, $F_0^{1,2}$, $F_0^{2,1}$, or $F_1^{0,1}$. In all but the last case, there is a unique pants decomposition of F, and, on $F = F_1^{0,1}$, we choose the pants decomposition containing the meridian of the torus. There is only one pair of pants associated with a basis for multiple arcs in F in each case, and (to simplify notation) we identify this pair of pants with the standard pants P. Identifying (the interior of) each component of $F - P$ with (the interior of) the standard annulus A once and for all, we say a train track $\tau \subset F$ with stops is *standard* if $\tau \cap P$ is either empty or one of the basic tracks in P and each component of $\tau \cap (F - P)$ is either empty or identified with a connector in A. In case $F = F_1^{0,1}$ where we assume that the boundary component of F is identified with the boundary component ∂_1 of P, we must furthermore rule out the *exceptional* possibility that $\tau \cap P$ is of archetype 2 or 3. It is straightforward to verify (as in Lemma 2.6.1) that each standard train track with stops is birecurrent.

Lemma 2.7.1: *Suppose that (τ, μ) is a positively measured generic train track with stops in a surface F of type $F_0^{0,3}$, $F_0^{1,2}$, $F_0^{2,1}$, or $F_1^{0,1}$. In case $F = F_1^{0,1}$, we furthermore assume that τ intersects the pants curve interior to F transversely in a single point (if at all). Then there is a standard positively measured train track with stops (τ', μ') in F so that (τ, μ) and (τ', μ') refine to a common measured train track.*

Proof: To begin, we assume that $F \neq F_1^{0,1}$ and construct the bifoliated neighborhood $(N, \mathcal{F}, \mathcal{F}^\perp)$ of τ in F associated to the positive measure μ as in Step 1 of Construction 1.7.7 (modified in §1.8 for train tracks with stops).

We claim that $\mathcal{F}$ has no semi-infinite singular leaves. Indeed, if there were such a singular leaf, then let ρ be the corresponding bi-infinite trainpath on τ. Since τ has only finitely many branches, ρ must contain a closed sub-trainpath corresponding to a simple closed curve γ in F. Since $\gamma < \tau$, the curve γ must be essential (as in §1.2), so γ must be homotopic to a boundary component of $P \subset F$ (since these are the only essential simple closed curves in F). It follows that ρ must terminate at a stop of τ, for otherwise, there would be a monogon (or bigon) immersed in F whose frontier corresponds to a trainpath (or to two trainpaths) in τ, and this contradicts Corollary 1.1.2 (in the setting of train tracks with stops).

Thus, the trainpath corresponding to a singular leaf of $\mathcal{F}$ lying in the interior of N is either a large trainpath (as in §2.3) or begins by traversing a large half-branch and ends at a stop of τ. We completely unzip along all the singular leaves of the former type as in §2.4. For a leaf of the latter type, we may unzip along a sub-arc of the leaf which terminates in a small neighborhood of the stop taking care to insure that sub-arcs corresponding to distinct leaves terminate on distinct ties of $\mathcal{F}^{\perp}$. These unzippings produce a bifoliated subset $N' \subset N$ as in §2.4. Collapsing each leaf of $N' \cap \mathcal{F}^{\perp}$ to a point produces a generic train track $\tau_1 \subset F$, and the measure $\mu > 0$ on τ induces a corresponding measure $\mu_1 > 0$ on τ_1. By construction, for each stop of τ_1, there is a neighborhood U about the boundary component of F containing it so that $\tau_1 \cap U$ either is empty or has the topological type of a tree whose root corresponds to the stop and which supports a smooth path from the root to each vertex of the tree (as in Figure 1.8.4b). Furthermore, each switch of τ_1 either lies in such a neighborhood or lies on a simple closed curve component of τ_1. Isotoping τ_1 in F, arrange that $\tau_1 \cap P$ consists of a collection of arcs properly imbedded in P and each component of $\tau_1 \cap (F - P)$ is empty, a closed curve component, or a tree as above. By Dehn's Theorem for multiple arcs in P (cf. §1.8), we may isotope $\tau_1 \cap P$ into canonical position.

Owing to the similarity of our basic tracks with the canonical models for multiple arcs in P, it is obvious that τ_1 is carried by some standard train track τ' with stops in F, and $\mu_1 > 0$ induces a measure $\mu' > 0$ on τ'. In particular, since τ' is transversely recurrent, so too is τ_1 by Lemma 1.3.3b (for tracks with stops). Thus, the hypotheses of Theorem 2.4.1 (for tracks with stops) are satisfied, and we conclude that (τ', μ') and (τ_1, μ_1) refine to a common measured train track (τ_2, μ_2).

Since (τ, μ) unzips to (τ_1, μ_1), we conclude that (τ, μ) refines to (τ_1, μ_1) by Lemma 2.4.2 (for track with stops). Thus, (τ, μ) also refines to (τ_2, μ_2) (by transitivity), as desired.

Turning finally to $F = F_1^{0,1}$, let $K \subset F$ denote the pants curve corresponding to the meridian of F, and identify a regular neighborhood V of K in F with the standard annulus. In case $\tau \cap V = \emptyset$, the lemma follows easily from the previous discussion by cutting on K. Thus, we may henceforth assume that $\tau \cap V$ consists of a single arc connecting the boundary components of V, and we let $V' \supset V$ be a slightly larger neighborhood with this same property. We also let V'' denote a small annular neighborhood of the boundary of F. The train track with stops $\tau \subset F$ gives rise to the train track with stops $\tau_1 = \tau \cap (F - V)$, and the positive measure μ on τ gives rise to a positive measure μ_1 on τ_1 in the natural way.

As in the previous argument, we identify $F - (V' \cup V'')$ with the pants P and refine (τ_1, μ_1) (by unzipping) to produce a train track with stops (τ_2, μ_2) so that $\tau_2 \cap P$ is a multiple arc in canonical position in P, each component of $\tau_2 \cap (V' - V)$ is a tree, as before, with its root lying in the boundary of V, and $\tau_2 \cap V''$ is empty, a closed curve component of τ_2, or a tree, as before, with its root lying in ∂F.

Let us now add V to $F - V$ extending (τ_2, μ_2) to a train track with stops (τ_3, μ_3) in F itself in the natural way, where $\tau_3 \cap V$ consists of a single arc connecting the roots of the trees. Furthermore, isotope a neighborhood of V' into V so that the resulting multiple arc $\tau_3 \cap P$ has vanishing twisting numbers (in Dehn's Theorem) associated with the boundary components of P isotopic to K; see Figure 2.7.1a. Finally, consider the bifoliated neighborhood associated to (τ_3, μ_3). For each switch of τ_3 lying in V, unzip along a sub-arc of the corresponding singular leaf, where the sub-arcs terminate inside V' and near $\partial V'$ to produce a generic measured train track (τ_4, μ_4) in F; see Figure 2.7.1b. Thus, $\tau_4 \cap V$ consists of a collection of arcs properly imbedded in V, and these arcs may be isotoped in V to canonical position.

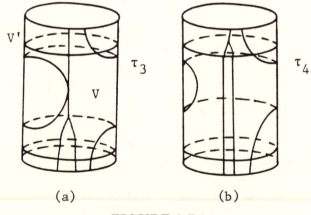

(a)   (b)

FIGURE 2.7.1

It is again obvious that there is a standard train track τ' in F which carries τ_4, and $\mu_4 > 0$ induces a measure $\mu' > 0$ on τ'. Arguing as before, we find that (τ, μ) and (τ', μ') have a common refinement, and the proof is complete.   q.e.d.

We are now in a position to prove the existence of the standard models.

Theorem 2.7.2: *Any measured train track* (τ, μ) *in* $F = F_g^s$ *is equivalent to a standard train track.*

Proof: Let $\mathcal{P}$ denote the pants decomposition of the basis $\mathcal{A}$. Applying Theorem 2.5.1, we may assume that $\mathcal{P}$ hits τ efficiently. Next, we alter (τ, μ) to arrange that τ intersects each pants curve in $\mathcal{P}$ in at most a single point. To do this, suppose that $K \in \mathcal{P}$ is a pants curve meeting τ in two points decomposing K into two arcs. We may suppose that one of these arcs, say $\alpha \subset K$, is free from other points of $K \cap \tau$, and there are two cases, as follows. If a trivial collapse of τ along α (as in §1.4) does not introduce a complementary bigon, then we perform this collapse. (Notice that such a collapse cannot introduce a complementary monogon since $\mathcal{P}$ hits τ efficiently.) If this collapse does introduce a bigon B, perform it anyway and let β be an arc in B connecting the vertices of the frontier of B; see Figure 2.7.2a. Collapsing B onto β as in Figure 2.7.2b (making choices) gives an equivalent measured train track (by Lemma 2.4.2 since the collapsed track unzips to the original track). In any case, we can decrease the number of times τ intersects components of $\mathcal{P}$, and this process terminates so that $\tau \cap K$ is at most a single point for each component K of $\mathcal{P}$.

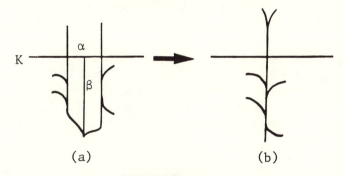

FIGURE 2.7.2

In case $F = F_1^1$, we may argue as in the last case of the previous lemma to produce an equivalent standard train track in F, proving the theorem in this case. We henceforth assume that $F \neq F_1^1$.

Let
$$\nu_i : A \to A_i, \quad i = 1, \ldots, N = 3g - 3 + s$$
$$f_j : P \to P_j, \quad j = 1, \ldots, M = 2g - 2 + s$$

be the characteristic maps for the basis $\mathcal{A}$. For each $j = 1, \ldots, M$, define the subsurface $P_j' \subset F$ to consist of P_j together with the annuli in the pants decomposition which meet P_j. Inspecting Figure 1.2.4, we find that each region P'_j is of type $F_0^{0,3}$, $F_0^{1,2}$, $F_0^{2,1}$, or $F_1^{0,1}$; indeed, the characteristic maps for $\mathcal{A}$ determine an identification of each P'_j with the corresponding surface, and we tacitly make these identifications.

Setting $j = 1$ and $(\tau_0, \mu_0) = (\tau, \mu)$, we proceed recursively, as follows. By the previous lemma, (τ_{j-1}, μ_{j-1}) is equivalent to a train track (τ_j, μ_j) so that $\tau_j \cap P'_j$ is a standard train track with stops in P'_j. Performing this construction for each $j = 1, \ldots, M$, we produce a sequence

$$(\tau, \mu) = (\tau_0, \mu_0), (\tau_1, \mu_1), \ldots, (\tau_M, \mu_M)$$

of equivalent measured train tracks in F.

By construction, $f_k^{-1}(\tau_K)$ is a basic track in P, and $\nu_i^{-1}(\tau_K)$ is a connector in A if $\bar{A}_i \cap \bar{P}_k \neq \emptyset$ for each $k \leq K$. It follows that τ_M is a standard train track in F, so (τ, μ) is equivalent to the standard train track (τ_M, μ_M), as desired.   q.e.d.

Using the result above, we next show that a recurrent non transversely recurrent train track arises from a birecurrent train track by excessive collapsing.

Corollary 2.7.3: *If (τ, μ) is a positively measured non transversely recurrent train track in F, then (τ, μ) splits to a birecurrent train track.*

Proof: By the previous theorem, (τ, μ) is equivalent to a positively measured standard train track (τ', μ') in F, and by Lemma 2.6.1, this standard train track is transversely recurrent. According to Theorem 2.3.1, (τ, μ) and (τ', μ') split to a common measured train track (τ'', μ''). Since $\tau' > \tau''$, the train track τ'' is transversely recurrent by Lemma 1.3.3b, and the proof is complete.   q.e.d.

Thus, if (τ, μ) is a positively measured train track, we may refine it to a positively measured transversely recurrent train track (τ', μ'). Construction 1.7.7 asssociates a measured geodesic lamination (G, λ) to the latter track, and we may attempt to extend Construction 1.7.7 to the setting of non transversely recurrent train tracks by associating (G, λ) to (τ, μ) itself. Among other things, our next result shows that this extension is well-defined.

Theorem 2.7.4: *Given a measure μ on the (not necessarily transversely recurrent) train track τ, the extension of Construction 1.7.7 above determines a well-defined measured geodesic lamination carried by τ. In fact,* *the extension gives rise to a bijection between $V(\tau)$ and the collection of measured geodesic laminations of compact support carried by τ, and (as in Theorem 1.7.12) the associated map $V(\tau) \to \mathcal{ML}_0(F)$ is continuous. Finally, equivalent measured train tracks give rise to the same measured geodesic lamination.*

Proof: Suppose that the positively measured train track (τ, μ) refines to (τ_i, μ_i), where τ_i is transversely recurrent, and let (G_i, λ_i) be the corresponding measured geodesic lamination, for $i = 1, 2$. Since (τ_1, μ_1) and (τ_2, μ_2) are equivalent, they are equivalent through birecurrent train tracks by Corollary 2.3.4. It follows from Proposition 2.2.3 that $(G_1, \lambda_1) = (G_2, \lambda_2)$, proving that our extension of Construction 1.7.7 is indeed well-defined. Since $G_1 < \tau_1 < \tau$, we have $G_1 < \tau$ by transitivity of carrying.

Next, we prove that the extension of Construction 1.7.7 determines a bijection between $V(\tau)$ and the collection of measured geodesic laminations carried by τ. Injectivity of the extension follows from Proposition 1.7.5. To see that the extension is surjective, suppose that (G, λ) is a measured geodesic lamination carried by τ inducing the measure μ, and let (G', λ') be the measured geodesic lamination associated to (τ, μ) by the extension. Of course, (G', λ') is also carried by τ and induces the same measure μ by construction. Since the extension is well-defined, as above, we conclude $(G, \lambda) = (G', \lambda')$ arises from (τ, μ), as desired.

To see that the associated map $V(\tau) \to \mathcal{ML}_0(F)$ is continuous, suppose that $\mu \in V(\tau)$ and (τ, μ) refines to (τ', μ') where τ' is transversely recurrent. Since Step 1 of Construction 1.7.7 did *not* require transverse recurrence, we may construct the measured laminations (G, λ) and (G', λ') corresponding to (τ, μ) and (τ', μ'), respectively. These two measured laminations are evidently isotopic, and continuity is easily proved as in Theorem 1.7.12.

For the final assertion, suppose that (τ, μ) and (τ', μ') are equivalent measured train tracks which refine to (τ_1, μ_1) and (τ_1', μ_1'), respectively, where τ_1 and τ_1' are birecurrent. Thus, (τ_1, μ_1) and (τ_1', μ_1') are equivalent birecurrent train tracks, so, as above, they determine the same measured geodesic lamination.   q.e.d.

Remark: In light of the previous result, we may identify $V(\tau)$ with the collection of measured geodesic laminations carried by τ for any train track τ. As mentioned in §2.1, it follows that if $\tau' < \tau$ with supporting map ϕ, then the incidence matrix $M^\phi : V(\tau') \to V(\tau)$ describes the inclusion $V(\tau') \subset V(\tau)$ for any choice of supporting map ϕ.

§2.8 UNIQUENESS OF THE STANDARD MODELS

Given a measured train track (τ, μ) in a surface F and a multiple curve $C \subset F$ hitting τ efficiently, we define the *variation of C with respect to (τ, μ)* by the formula

$$\mathcal{V}_{(\tau,\mu)}(C) = \sum n_i \mu(b_i),$$

where the sum is over all branches b_i of τ and n_i denotes the number of times C meets b_i. Thus, the variation function detects the amount that C crosses the track τ, where each crossing of branch b_i is weighted by $\mu(b_i)$.

We first show that $\mathcal{V}_{(\tau,\mu)}(C)$ is invariant under isotopy of C. To this end, let C_0 and C_1 be representatives of the same multicurve and suppose that both C_0 and C_1 hit τ efficiently. We assume that C_0 and C_1 are each connected closed curves; the general case is similar. To begin, suppose that C_0 and C_1 are disjoint, let A be the annulus imbedded in F with boundary $C_0 \cup C_1$, and choose a (generic) product structure $S^1 \times [0,1]$ on A. Since C_0 and C_1 each hit τ efficiently, we may alter τ by an isotopy in A fixing ∂A so that each $S^1 \times \{p\}$, for $p \in [0,1]$, hits τ efficiently, except for a finite collection $p_1 < p_2, \ldots < p_K$ of points in $[0,1]$ where $\tau \cap (S^1 \times \{p_k\})$ contains a single switch of τ. The switch conditions then show that $\mathcal{V}_{(\tau,\mu)}(S^1 \times \{p_k \pm \epsilon\})$ is constant for small ϵ, and it follows that $\mathcal{V}_{(\tau,\mu)}(C_0) = \mathcal{V}_{(\tau,\mu)}(C_1)$.

If C_0 and C_1 are not disjoint, then a standard argument shows that there is a sequence $C_0 = D_0, D_1, \ldots, D_n = C_1$ of isotopic simple closed curves so that there is exactly one bigon B_i with imbedded closure complementary to $D_i \cup D_{i+1}$ for $i = 0, \ldots, n - 1$; moreover, one can insure that each D_i hits τ efficiently. Consider the curve D obtained from D_i by isotoping across B_i and slightly across D_{i+1}; the curve D is disjoint from both D_i and D_{i+1} (and can be perturbed to hit τ efficiently), so by the argument above,

$$\mathcal{V}_{(\tau,\mu)}(D_i) = \mathcal{V}_{(\tau,\mu)}(D) = \mathcal{V}_{(\tau,\mu)}(D_{i+1}),$$

and our claim is proved.

Recall that $[\tau, \mu]$ denotes the equivalence class (under splitting, shifting, and isotopy) of the measured train track (τ, μ), and define the *variation function*

$$\mathcal{V}_{[\tau,\mu]} : \mathcal{S}(F) \to \left[\mathbb{R}_+ \cup \{0\} \right]$$

of $[\tau, \mu]$ as follows. Given a multicurve $c \in \mathcal{S}(F)$, we may choose representatives $C' \in c$ and $(\tau', \mu') \in [\tau, \mu]$ so that C' hits τ' efficiently by Theorem 2.5.1 and define

$$\mathcal{V}_{[\tau,\mu]}(c) = \mathcal{V}_{(\tau',\mu')}(C').$$

Proposition 2.8.1: *The variation function of an equivalence class of measured train tracks is well-defined.*

Proof: Suppose that (τ_1, μ_1) and (τ_2, μ_2) are equivalent positively measured train tracks and C_1 and C_2 are isotopic multiple curves, where C_i hits τ_i efficiently, for $i = 1, 2$. By Theorem 2.3.1, (τ_1, μ_1) and (τ_2, μ_2) split to a common track (τ, μ); we shall show that C_i is isotopic to some multiple curve C_i' which hits τ efficiently and $\mathcal{V}_{(\tau,\mu)}(C_i') = \mathcal{V}_{(\tau_i,\mu_i)}(C_i)$, for $i = 1, 2$. Thus,

$$\mathcal{V}_{(\tau_1,\mu_1)}(C_1) = \mathcal{V}_{(\tau,\mu)}(C_1') = \mathcal{V}_{(\tau,\mu)}(C_2') = \mathcal{V}_{(\tau_2,\mu_2)}(C_2),$$

where the second equality follows from the previous remarks.

Concentrating on our assertion for C_1 and (τ_1, μ_1), let $(N, \mathcal{F}, \mathcal{F}^\perp)$ be the bifoliated neighborhood associated to (τ_1, μ_1) in Step 1 of Construction 1.7.7. Isotope C_1 to a multiple curve C_1' so that $C_1' \cap N$ consists of ties of $\mathcal{F}^\perp$ and C_1' hits τ_1 efficiently. Perform the splits from (τ_1, μ_1) to (τ, μ) so that all the intermediary train tracks lie inside N transverse to $\mathcal{F}^\perp$, and suppose that (τ', μ') is one of the intermediary train tracks. A bigon with frontier in $C_1' \cup \tau'$ must be interior to N since C_1' hits τ_1 efficiently, and this is impossible since $C_1' \cap N$ consists of ties of $\mathcal{F}^\perp$ and τ' is transverse to $\mathcal{F}^\perp$. Thus, C_1' hits each intermediary track efficiently and, in particular, hits τ efficiently.

Using the definition of measures induced by splitting, one easily proves that $\mathcal{V}_{(\tau',\mu')}(C_1') = \mathcal{V}_{(\tau_1,\mu_1)}(C_1)$, for each intermediary track, so, in particular $\mathcal{V}_{(\tau,\mu)}(C_1') = \mathcal{V}_{(\tau_1,\mu_1)}(C_1)$, as desired.   q.e.d.

The following theorem is our main result for this section and has numerous consequences in the sequel.

Theorem 2.8.2: *Distinct standard train tracks in a fixed surface determine distinct variation functions.*

Proof: Consider the surface $F = F_g^s$ and let

$$f_j : P \to P_j, \quad j = 1, \ldots, M = 2g - 2 + s$$
$$\nu_i : A \to A_i, \quad i = 1, \ldots, N = 3g - 3 + s$$

be the characteristic maps for the pairs of pants and annuli, respectively, in the standard basis $\mathcal{A}$ for multicurves on a surface F. In §2.6, a complex $\mathcal{K}(F)$ of standard tracks in F was defined and a continuous injection

$$q : \mathcal{K}(F) \to \mathbb{R}^{6M+2N}$$

was described. We adopt the notation

$$\lambda_{mn}^j, \quad \text{for } 1 \leq m \leq n \leq 3, \quad j = 1, \ldots, M$$
$$\epsilon_t^i, \quad \text{for } t = 1, 2, \quad i = 1, \ldots, N$$

for the basic parameters as in §2.6 and prove the theorem by computing $q((\tau, \mu))$ from $\mathcal{V}_{[\tau, \mu]}$ for $\tau \subset F$ a standard train track. In fact, we shall give a finite set of connected multiple curves so that the variations of the corresponding multicurves uniquely determine the basic parameters.

To begin, let

$$m_i = m_i(\tau, \mu) = \mathcal{V}_{[\tau, \mu]}([K_i]), \quad \text{for } i = 1, \ldots, N,$$

where K_i is a pants curve in the basis $\mathcal{A}$ with corresponding multiple curve $[K_i]$. As in the proof of Proposition 2.6.3, these intersection numbers uniquely determine the parameters $\{\lambda_{mn}^j\}$. Thus, we may regard these parameters as known in terms of the variation function.

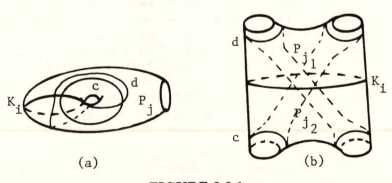

(a)   (b)

FIGURE 2.8.1

We must compute the parameters $\{\epsilon_t^i\}$, for $t = 1, 2$ and $i = 1, \ldots, N$, from the variation function and begin with the (essentially general) case in which F is closed ($s = 0$). In this case, each pants curve K_i in $\mathcal{A}$ appears in one of the configurations depicted in Figure 2.8.1a and 2.8.1b; namely, if $\mathcal{P}$ is the pants decomposition corresponding to $\mathcal{A}$, then the component of $(F - \mathcal{P}) \cup K_i$ containing K_i is a surface of type $F_1^{0,1}$ or $F_0^{0,4}$, respectively. We henceforth refer to these respective possibilities as case (a) and case (b) and adopt the notation indicated in Figure 2.8.1 for nearby pairs of pants in each case; also indicated in Figure 2.8.1 are a pair c and d of connected multicurves. Figure 2.8.2 indicates the basis for multiarcs on each subsurface induced from the basis $\mathcal{A}$ for multicurves on F itself.

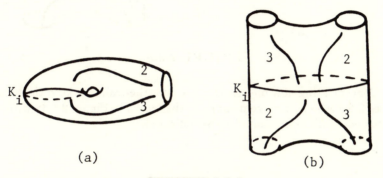

(a)   (b)

FIGURE 2.8.2

We shall split each standard train track sufficiently until it hits both c and d efficiently and compute (from a picture) the quantities

$$\mathcal{V}_c = \mathcal{V}_{[\tau,\mu]}(c) \quad \text{and} \quad \mathcal{V}_d = \mathcal{V}_{[\tau,\mu]}(d)$$

in terms of the basic parameters with respect to the bases of Figure 2.8.2. Finally, we must invert the formulas derived to solve for $\{\epsilon_t^i\}$ from $\mathcal{V}_c$, $\mathcal{V}_d$, and the known parameters $\{\lambda_{mn}^j\}$.

To this end, we first observe that in case (a), $\mathcal{V}_c$ and $\mathcal{V}_d$ are linear in the known parameter λ_{11}^j, so we may assume that τ meets the subsurface in a subtrack of the train tracks with stops indicated in Figure 2.8.3a, where the annulus A_i may contain any one of the connectors. In the same way, in case (b), $\mathcal{V}_c$ and $\mathcal{V}_d$ are linear in the known basic parameters

$$\{\lambda_{mn}^j : 2 \le m \le n \le 3 \text{ and } j = j_1, j_2\},$$

so we may assume that τ meets the subsurface in a subtrack of the various train tracks with stops indicated in Figure 2.8.3b, where the connector is again arbitrary.

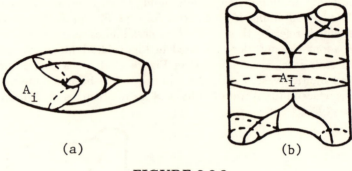

(a)   (b)

FIGURE 2.8.3

Concentrating now on case (a), consider the train tracks with stops in Figure 2.8.4 which arise from splitting the tracks in Figure 2.8.3a with the various possibilities of connector, as indicated.

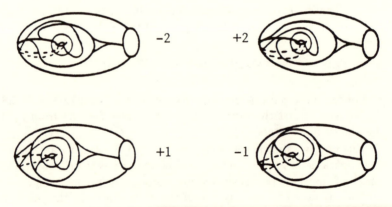

FIGURE 2.8.4

We compute $\mathcal{V}_c$ and $\mathcal{V}_d$ in each case of Figure 2.8.4 in terms of

$$\epsilon_t = \epsilon_t^i, \text{ for } t = 1, 2, \text{ and } m = m_i,$$

and our results are summarized in the following table.

type	variations
$+2$	$\mathcal{V}_c = \varepsilon_1 + \varepsilon_2$ $\mathcal{V}_d = \varepsilon_2$
$+1$	$\mathcal{V}_c = \varepsilon_1$ $\mathcal{V}_d = \varepsilon_2$
-1	$\mathcal{V}_c = -\varepsilon_1$ $\mathcal{V}_d = \varepsilon_2 - 2\varepsilon_1$
-2	$\mathcal{V}_c = \varepsilon_2 - \varepsilon_1$ $\mathcal{V}_d = \varepsilon_2 - 2\varepsilon_1$

VARIATIONS IN CASE (a)

To determine the type of a maximal connector containing the given connector from this data, we first compute directly from the table above that

if the type is $+2$, then $\mathcal{V}_c - \mathcal{V}_d = m$,

if the type is $+1$, then $-m \leq \mathcal{V}_c - \mathcal{V}_d \leq m$,

if the type is -1, then $\mathcal{V}_c - \mathcal{V}_d = -m$,

if the type is -2, then $\mathcal{V}_c - \mathcal{V}_d = -m$,

and, furthermore, the subtracks of a maximal connector of type $+1$ correspond to the cases of equality above. Thus, the connector is maximal of type $+1$ if and only if $|\mathcal{V}_c - \mathcal{V}_d| < m$, and it is a subtrack of type $+2$ if and only if $\mathcal{V}_c - \mathcal{V}_d = m$.

It remains to distinguish between types -1 and -2 in case $\mathcal{V}_c - \mathcal{V}_d = -m$. To this end, we compute directly from the previous table that

if $\mathcal{V}_c \geq |m - \mathcal{V}_d|$, then the type is -2,

if $\mathcal{V}_c = |m - \mathcal{V}_d|$, then the type is -1,

and, furthermore, the common subtrack of types -1 and -2 corresponds to equality in the former case.

Thus, the type of a maximal connector containing the given connector can be determined from the variation function, as above.

To complete our discussion of case (a), it remains to give formulas for the basic parameters ε_1 and ε_2, and, since the type of a superconnector is known, this is easily done. Again checking directly, one finds that if the connector is a subtrack of a maximal connector of type +2 or -2, then

$$\epsilon_1 = \mathcal{V}_c - \mathcal{V}_d$$
$$\epsilon_2 = \mathcal{V}_c - |\mathcal{V}_c - \mathcal{V}_d|,$$

while if the type is +1 or -1, then

$$\epsilon_1 = m_i - \mathcal{V}_d$$
$$\epsilon_2 = m_i - \mathcal{V}_c.$$

The basic parameters may therefore be computed in terms of the quantities $\mathcal{V}_c$, $\mathcal{V}_d$, and m, completing our discussion of case (a).

Turning to case (b), consider the train tracks of Figure 2.8.5 which arise from splitting the tracks with stops in Figure 2.8.3b with the various possibilities for connector.

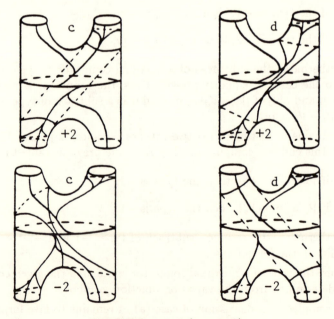

FIGURE 2.8.5 (type ±2)

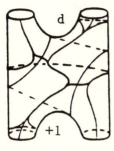
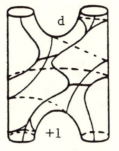
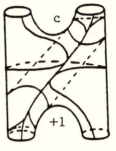
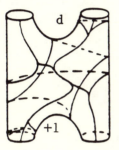
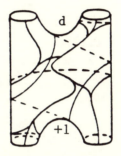

FIGURE 2.8.5 (type +1)

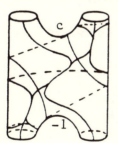
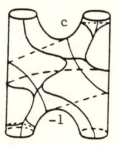
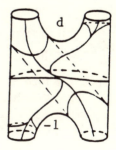
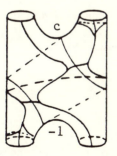
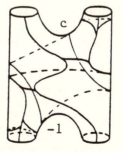

FIGURE 2.8.5 (type −1)

As before, we compute V_c and V_d from the basic parameters

$$m = m_i, \epsilon_t = \epsilon_t^i, \text{ for } t = 1, 2,$$
$$\lambda_{mn} = \lambda_{mn}^{j_1}, \text{ and } \kappa_{mn} = \lambda_{mn}^{j_2}, \text{ for } 1 \leq m \leq n \leq 3$$

in each case of the figure (where the superscripts j_1 and j_2 refer to the pants P_{j_1} and P_{j_2} in Figure 2.8.1b). The results are summarized in the following table.

type	variations
$+2$	$V_c = 2(\varepsilon_1 + \varepsilon_2) + \lambda_{12} + 2\lambda_{11} + \kappa_{12} + 2\kappa_{11}$ $V_d = 2\varepsilon_2 + \lambda_{12} + 2\lambda_{11} + \kappa_{12} + 2\kappa_{11}$
$+1$	$V_c = 2\varepsilon_1 + \lambda_{12} + 2\lambda_{11} + \kappa_{12} + 2\kappa_{11}$ $V_d = |\varepsilon_1 - \kappa_{13}| + |\varepsilon_1 - \lambda_{13}|$
-1	$V_c = |\varepsilon_2 - \lambda_{13}| + |\varepsilon_2 - \kappa_{13}|$ $V_d = -2\varepsilon_1 + \lambda_{13} + \kappa_{13}$
-2	$V_c = 2\varepsilon_2 + \lambda_{13} + \kappa_{13}$ $V_d = 2(\varepsilon_2 - \varepsilon_1) + \lambda_{13} + \kappa_{13}$

VARIATIONS IN CASE (b)

To determine the type of the connector from this data, we first compute directly from the previous table that

if the type is $+2$, then $V_c - V_d = 2m$.

if the type is $+1$, then $-2m \leq V_c - V_d \leq -2m$,

if the type is -1, then $V_c - V_d \leq -2m$,

if the type is -2, then $V_c - V_d = -2m$.

Thus, the type is -1 if $V_c - V_d < -2m$, and the type is $+1$ if $-2m < V_c - V_d < 2m$. It remains therefore only to distinguish between the types in case $V_c - V_d = \pm 2m$.

If $V_c - V_d = 2m$, then direct computation shows that

if $V_c + \lambda_{13} + \kappa_{13} \geq 4m$, then the type is $+2$,

if $V_c + \lambda_{13} + \kappa_{13} \leq 4m$, then the type is $+1$,

and equality holds on the common subtrack. Thus, we may distinguish between types $+1$ and $+2$ if $V_c - V_d = 2m$.

If $V_c - V_d = -2m$, then direct computation shows that the type is $+1$ if $V_d = \lambda_{13} + \kappa_{13}$ (while $V_d \geq \lambda_{13} + \kappa_{13}$ in the remaining cases), and equality holds on the common subtrack of types -1 and $+1$. It remains only to distinguish between types -1 and -2. To this end, first observe that for type -1, we must have $\epsilon_2 \geq \lambda_{13}$ and $\epsilon_2 \geq \kappa_{13}$ (since we are assuming that $V_c - V_d = -2m$). It follows that

if $V_c \leq \lambda_{13} + \kappa_{13}$, then the type is -1,

if $V_c \geq \lambda_{13} + \kappa_{13}$, then the type is -2,

and equality holds on the common subtrack. Thus, we may distinguish between types ± 1 and -2 in case $V_c - V_d = -2m$.

To complete our discussion of case (b), we finally give formulas for the basic parameters in terms of the variations. If the connector is a subtrack of a maximal connector of type $+2$ or -2, then one finds that

$$2\epsilon_1 = V_c - V_d,$$

$$2\epsilon_2 = \begin{cases} V_d - (\lambda_{12} + 2\lambda_{11} + \kappa_{12} + 2\kappa_{11}), & \text{if the type is } +2; \\ V_c - \lambda_{13} - \kappa_{13}, & \text{if the type is -2,} \end{cases}$$

while for a connector which is a subtrack of a maximal connector of type $+1$ or -1, we find

$$2\epsilon_1 = \begin{cases} V_c - (\lambda_{12} + 2\lambda_{11} + \kappa_{12} + 2\kappa_{11}), & \text{if the type is } +1; \\ -V_d + \lambda_{13} + \kappa_{13}, & \text{if the type is -1,} \end{cases}$$

$$\epsilon_2 = m_i - |\epsilon_1|.$$

We have thus succeeded in computing the coordinate functions of the injection q from the values of the variation function for a standard train track in case F is closed, as desired. To complete the proof of the theorem, we must consider the possibility that F is punctured. If K is a pants curve in the pants decomposition $\mathcal{P}$ associated to the standard basis $\mathcal{A}$ on F, then inspection of Figure 2.5.1 shows that the component R of $(F - \mathcal{P}) \cup K$ containing K is either a surface of type $F_1^{s',r'}$, where $s' + r' = 1$, or a surface of type $F_0^{s',r'}$, where $s' + r' = 4$ and $r' \geq 1$. We consider curves in R analogous to the curves c, d used above and observe that the computations given before apply verbatim to allow the computation of basic parameters in R from the variation function of a standard track in F. In effect, the computations on R amount to the application of the formulas above to a subtrack of one of the train tracks with stops already considered.

The general case therefore follows immediately from the previous computations, and the proof is complete.   q.e.d.

We have found that in the surface $F = F_g^s$, there are connected multiple curves $\{c_k\}_1^{3N}$ whose corresponding variations $V_k = V_{[\tau,\mu]}(c_k)$, for $k =$

$1, \ldots, 3N$, uniquely determine the equivalence class of a standard measured train track (τ, μ) in F. As an addendum to the proof of Theorem 2.8.2, we have

Corollary 2.8.3: *The coordinate functions of the homeomorphism*

$$\pi : \mathcal{K}(F) \to \mathrm{IR}^{2N}$$

(defined in §2.6) are continuous functions of the variations $\{\mathcal{V}_k\}_1^{3N}$.

Proof: Insofar as each Dehn-Thurston intersection number is itself one of the specified variations, continuity of these intersection numbers is immediate. Furthermore, one checks directly that the piecewise-continuous forumulas for ϵ_1^i and ϵ_2^i given in the proof of Theorem 2.8.2 yield a Dehn-Thurston twisting number (computed from ϵ_1^i and ϵ_2^i as in §2.6) which extends continuously across each face other than that corresponding to the common face of types $+2$ and -2. For instance, adopting the notation of case (a) in the proof of the previous theorem, the twisting number is

$$t_i = \begin{cases} -|\epsilon_1^i| = -\mathcal{V}_c, & \text{if the type is -1;} \\ +|\epsilon_1^i| = +\mathcal{V}_c, & \text{if the type is } +1. \end{cases}$$

On the common face which corresponds to a subtrack of a connector of type -1 and of type $+1$, we find that $\mathcal{V}_c = 0$, so t_i indeed extends continuously across this face. We leave the routine (but somewhat messier) verifications of continuity across the other faces as an exercise for the untiring reader.

Finally, one checks that the coordinate functions of π are derived from the intersection and twisting numbers in such a way that they extend continuously across the common faces of types $+2$ and -2 as well, completing the proof.   q.e.d.

Pulling together various results, we have

Theorem 2.8.4: *Any measured train track is equivalent to a unique standard measured train track.*

Proof: Any measured train track (τ, μ) is equivalent to some standard measured train track (τ', μ') by Theorem 2.7.2. Proposition 2.8.1 shows that $\mathcal{V}_{[\tau, \mu]} = \mathcal{V}_{[\tau', \mu']}$, and Theorem 2.8.2 finally shows that $V_{[\tau', \mu']}$ uniquely determines the standard train track (τ', μ').   q.e.d.

If (G, λ) is a measured geodesic lamination in F, then there is a "variation function"

$$\mathcal{V}_{(G, \lambda)} : \mathcal{S}(F) \to [\mathrm{IR}_+ \cup \{0\}]$$

defined as follows. If $c \in \mathcal{S}(F)$, then we may represent c by a geodesic C (in the same hyperbolic structure for which G is geodesic) and define

$$\mathcal{V}_{(G,\lambda)}(c) = \int_C \lambda;$$

components of C which are leaves of G (if any) do not contribute to the integral by definition. Thus, $\mathcal{V}_{(G,\lambda)}(c)$ is the infimum of the λ-transverse measure of all multiple curves representing c.

It is clear from the definitions that if (τ, μ) is a positively measured transversely recurrent train track in F giving rise to (G, λ) via Construction 1.7.7, then $\mathcal{V}_{[\tau,\mu]} = \mathcal{V}_{(G,\lambda)}$ as functions on $\mathcal{S}(F)$. Furthermore, dropping the restriction in the previous sentence that τ is transversely recurrent and using the extension of Construction 1.7.7 in Theorem 2.7.4, the equality $\mathcal{V}_{[\tau,\mu]} = \mathcal{V}_{(G,\lambda)}$ follows directly from Proposition 2.8.1.

Our next result shows that two measured train tracks are equivalent if and only if they give rise to the same measured geodesic laminations.

Theorem 2.8.5: *Suppose that (τ_i, μ_i) are measured train tracks in F giving rise to the measured geodesic laminations (G_i, λ_i), respectively, by the extension of Construction 1.7.7. The following are equivalent*

(i) *The measured train tracks (τ_1, μ_1) and (τ_2, μ_2) are equivalent.*

(ii) *The variation functions $\mathcal{V}_{[\tau_1,\mu_1]}$ and $\mathcal{V}_{[\tau_2,\mu_2]}$ are identical.*

(iii) *The geodesic laminations (G_1, λ_1) and (G_2, λ_2) are identical.*

Proof: By Proposition 2.8.1, if two measured train train tracks are equivalent, then they have the same variation function. Conversely, if two measured train tracks have the same variation function, then so too do the standard tracks equivalent to them, again by Proposition 2.8.1. These standard tracks are therefore identical by Theorem 2.8.2, so the original measured tracks are equivalent. We have therefore shown that (i) is equivalent to (ii). That (i) implies (iii) is the content of the last part of Theorem 2.7.4. To see that (iii) implies (ii), it suffices to show that distinct standard measured train tracks give rise to distinct measured geodesic laminations, and this follows immediately from Theorem 2.8.2 and the remarks above on the variation function of a measured geodesic lamination. q.e.d.

In light of the previous result, one may identify an equivalence class of measured train tracks with its corresponding measured geodesic lamination.

Recall the space $\mathcal{K}(F)$ of standard measured train tracks in F and observe that Construction 1.7.7 induces a canonical map

$$\kappa : \mathcal{K}(F) \to \mathcal{ML}_0(F).$$

We have

Corollary 2.8.6: *The map κ is a homeomorphism.*

Proof: Bijectivity of κ follows immediately from Theorems 2.8.2 and 2.8.5. Insofar as $\mathcal{K}(F)$ is the quotient of the closed sets $V(\tau)$ for $\tau \subset F$ standard, κ is continuous by Theorem 1.7.12. Corollary 2.8.3 shows that κ^{-1} is continuous, and the proof is complete.   q.e.d.

Consequences of the previous result and a further investigation of the topology of $\mathcal{ML}_0(F)$ will be pursued in the next chapter.

CHAPTER 3   THE STRUCTURE OF $\mathcal{ML}_0$

This chapter is dedicated to a further study of the structure of $\mathcal{ML}_0(F)$, and we begin with the Dehn-Thurston Theorem (from [P1]), which gives global coordinates on this space. This result is used to show that $\mathcal{ML}_0(F)$ is a topological open ball and the associated projective space $\mathcal{PL}_0$ is a topological sphere. A complete train track in F corresponds to a chart on these manifolds, and the manifold structure on $\mathcal{ML}_0(F)$ is found to be somewhat richer than simply piecewise-linear. The collection $\mathcal{S}(F)$ of multicurves in F canonically imbeds in $\mathcal{ML}_0(F)$ and in $\mathcal{PL}_0(F)$, and we show that the collection of connected multicurves is dense in $\mathcal{PL}_0(F)$. Thus, $\mathcal{PL}_0(F)$ may be thought of as a spherical completion of the discrete set $\mathcal{S}(F)$. We also show that homology intersection numbers of real cycles in F induce a natural symplectic structure on the piecewise-linear manifold $\mathcal{ML}_0(F)$. Next, we consider an equivalence relation on $\mathcal{ML}_0(F)$, where two measured geodesic laminations are regarded as equivalent if the underlying (unmeasured) geodesic laminations coincide, and we give a reasonably explicit description of the associated partition of $\mathcal{ML}_0(F)$. Finally, we explore a deep and beautiful duality between transverse and tangential structures on a certain generalization of train tracks.

§3.1 THE TOPOLOGY OF $\mathcal{ML}_0$ AND $\mathcal{PL}_0$

In this section, we derive several basic results on the topology of $\mathcal{ML}_0$ and $\mathcal{PL}_0$. To a large extent, these results follow more or less directly from previous considerations, and, in particular, the standard tracks are seen to play a central role in our treatment. The train tracks we consider throughout this chapter will be assumed to be generic and birecurrent unless otherwise indicated. (In light of Theorem 2.7.4, transverse recurrence is not essential to our discussions here, but we make this assumption for convenience.) Furthermore, we again concentrate on train tracks without stops in surfaces without boundary leaving the extensions of our results to bounded surfaces and to train tracks with stops as exercises for the reader. To begin, we have

Theorem 3.1.1 [The Dehn-Thurston Theorem]: *Fix any surface $F = F_g^s$ and let $N = 3g - 3 + s$. The map*

$$\mathcal{ML}_0(F) \to \mathbb{R}^{2N}$$
$$(G, \lambda) \mapsto \big(m_1(\tau, \mu), t_1(\tau, \mu), \ldots m_N(\tau, \mu), t_N(\tau, \mu)\big)$$

gives global coordinates on $\mathcal{ML}_0(F)$, where (τ, μ) is the (unique) standard measured train track representing (G, λ), and m_i and t_i denote the intersection number and twisting number, respectively, with respect to the ith pants curve of the standard basis for multicurves on F (as in §2.6).

Proof: According to Corollary 2.8.6, $\mathcal{ML}_0(F)$ is homeomorphic to the complex $\mathcal{K}(F)$ of standard measured tracks in F. It follows from Proposition 2.6.3 that the intersection and twisting numbers give global coordinates on $\mathcal{K}(F)$, and the result follows. q.e.d.

Thus, given a measured geodesic lamination in F, there is a unique corresponding standard measured train track in F, and we refer to the intersection and twisting numbers of this standard measured train track as the intersection and twisting numbers of the measured geodesic lamination

174

itself. We refer to these parameters taken together as the *Dehn-Thurston coordinates* on $\mathcal{ML}_0(F)$, and these are global coordinates on $\mathcal{ML}_0(F)$.

To analyze the global topology of $\mathcal{ML}_0(F)$ and $\mathcal{PL}_0(F)$, we again apply Corollary 2.8.6 and Proposition 2.6.3 to conclude that $\mathcal{ML}_0(F)$ is homeomorphic to $\mathbb{R}^{2N}$, where $N = 3g-3+s$. Since $\mathcal{ML}_0(F)$ is homeomorphic to the cone over $\mathcal{PL}_0(F)$ from the empty lamination, we furthermore conclude that $\mathcal{PL}_0(F)$ is homeomorphic to a topological sphere of dimension $2N - 1$. The observation that the Dehn-Thurston theorem itself proves that $\mathcal{PL}_0(F)$ is a topological sphere was pointed out by Allen Hatcher.

As in §2.2, since $\tau \subset F$ is transversely recurrent, we may identify the cone $V(\tau)$ of measures on τ with the corresponding subset of $\mathcal{ML}_0(F)$ via Construction 1.7.7, and we similarly identify the polyhedron $U(\tau)$ with the corresponding subset of $\mathcal{PL}_0(F)$. We furthermore let $\overset{\circ}{V}(\tau)$ denote the collection of all strictly positive measures and $\overset{\circ}{U}(\tau)$ denote the collection of all strictly positive projective measures on τ identifying these open balls with the corresponding subsets of $\mathcal{ML}_0(F)$ and $\mathcal{PL}_0(F)$, respectively.

Lemma 3.1.2: *If $\tau \subset F$ is a complete train track, then $\overset{\circ}{V}(\tau)$ is open in $\mathcal{ML}_0(F)$, and $\overset{\circ}{U}(\tau)$ is open in $\mathcal{PL}_0(F)$.*

Proof: Suppose first that $F \neq F_1^1$ so the complete train track $\tau \subset F$ is non-orientable. According to Lemma 2.1.1 and Corollary 1.1.3, $\overset{\circ}{V}(\tau)$ is an open ball of dimension $3\chi(F)-s = 6g-6+2s = 2N$ imbedded in $\mathcal{ML}_0(F)$, which is a topological ball of dimension $2N$ by the remarks above. Openness of $\overset{\circ}{V}(\tau) \subset \mathcal{ML}_0(F)$ therefore follows from invariance of domain, and openness of $\overset{\circ}{U}(\tau) \subset \mathcal{PL}_0(F)$ follows directly from this.

For the surface $F = F_1^1$, each of $\mathcal{ML}_0(F)$ and $\overset{\circ}{V}(\tau)$ are open balls of dimension 2, and the lemma follows, as above, in this case as well. q.e.d.

In light of this result, one often thinks of a complete train track $\tau \subset F$ as a chart on the manifold $\mathcal{ML}_0(F)$ or $\mathcal{PL}_0(F)$. Given a point $(G, \lambda) \in \mathcal{ML}_0(F)$, one can construct such a chart about it as follows. By Corollary 1.7.6, G is carried by some generic birecurrent track $\tau \subset F$, and we may assume that the corresponding measure μ on τ is positive. Of course, τ may not be complete, but we can perform trivial collapses along admissible arcs to produce a complete train track $\tau' \subset F$ by Proposition 1.4.9. The measure $\mu > 0$ on τ induces a measure $\mu' > 0$ on τ', and $\overset{\circ}{V}(\tau')$ is therefore a chart about (G, λ) in $\mathcal{ML}_0(F)$, as desired.

Next, we investigate the relationship between the space $\mathcal{ML}_0(F)$ and the (discrete) set $\mathcal{S}(F)$ of multicurves in F. To this end, suppose that

$C \in \mathcal{S}(F)$ consists of components C_j^k, where $j = 1, \ldots, J, k = 1, \ldots, n_j \geq 1$, and C_j^k is isotopic to $C_{j'}^{k'}$ if and only if $j = j'$. Straighten $\{C_j^1\}_{j=1}^J$ to geodesics (for some hyperbolic structure on F) and take the transverse (counting) measure on the leaf C_j^1 to be n_j, for each $j = 1, \ldots, J$. This defines a measured geodesic lamination $G_C \subset F$ corresponding to $C \in \mathcal{S}(F)$ and determines a natural imbedding

$$\mathcal{S}(F) \to \mathcal{ML}_0(F)$$
$$C \mapsto G_C.$$

Identifying $\mathcal{S}(F)$ with $\mathcal{S}(F) \times \{1\}$, this extends to an imbedding

$$[\mathcal{S}(F) \times \mathbb{R}_+] \to \mathcal{ML}_0(F)$$
$$C \times t \mapsto tG_C,$$

where if G denotes the measured geodesic lamination (G, λ), then tG denotes the measured geodesic lamination $(G, t\lambda)$.

Using these imbeddings, we henceforth regard

$$\mathcal{S}(F) \subset [\mathcal{S}(F) \times \mathbb{R}_+] \subset \mathcal{ML}_0(F).$$

In particular, we may restrict to the subset $\mathcal{S}'(F) \subset \mathcal{S}(F)$ (which, we recall, consists of *connected* multicurves) and regard

$$\mathcal{S}'(F) \subset [\mathcal{S}'(F) \times \mathbb{R}_+] \subset \mathcal{ML}_0(F).$$

Furthermore, the natural $\mathbb{R}_+$-action on $\mathcal{ML}_0(F)$ restricts to the multiplicative action of $\mathbb{R}_+$ on the second factor of $\mathcal{S}(F) \times \mathbb{R}_+$ and determines canonical imbeddings

$$\mathcal{S}'(F) \subset \mathcal{S}(F) \subset \mathcal{PL}_0(F).$$

Theorem 3.1.3: *For any surface F, $\mathcal{S}'(F) \times \mathbb{R}_+$ is dense in $\mathcal{ML}_0(F)$, and $\mathcal{S}'(F)$ is dense in $\mathcal{PL}_0(F)$.*

Proof: The second assertion evidently follows from the first, so we concentrate on proving that $\mathcal{S}'(F) \times \mathbb{R}_+$ is dense in $\mathcal{ML}_0(F)$.

We first show that $\mathcal{S}(F) \times \mathbb{R}_+$ is dense in $\mathcal{ML}_0(F)$. As above, a point in $\mathcal{ML}_0(F)$ corresponds to a positively measured complete train track (τ, μ), and, by Lemma 3.1.2, it suffices to exhibit a sequence in $\overset{\circ}{V}(\tau) \cap [\mathcal{S}(F) \times \mathbb{R}_+]$ tending to μ in $V(\tau)$. By recurrence of τ, there is a sequence $\mu_i \in \overset{\circ}{V}(\tau)$ taking rational values on each branch of τ, for $i \geq 1$. Clearing denominators for each i, there is a natural number N_i so that $N_i\mu_i \in \overset{\circ}{V}(\tau)$ is an integral measure on τ, and we let $C_i \subset F$ denote the corresponding multicurve. The sequence $N_i^{-1}G_{C_i}$ for $i \geq 1$ evidently lies in $\overset{\circ}{V}(\tau) \cap [\mathcal{S}(F) \times \mathbb{R}_+]$ and converges to μ, as desired.

In case $F = F_1^1$, $\mathcal{S}'(F) = \mathcal{S}(F)$, and the proof is complete, so we henceforth assume that $F \neq F_1^1$. Using a standard diagonal argument, it remains to show that if $C \subset F$ is a multicurve, then $G_C \in \mathcal{ML}_0(F)$ is the limit of a sequence in $\mathcal{S}'(F) \times \mathbb{R}_+$.

To prove the claim, suppose that C consists of components C_j^k, for $j = 1, \ldots, J$ and $k = 1, \ldots, n_j \geq 1$, where C_j^k is isotopic to $C_{j'}^{k'}$ if and only if $j = j'$, as above. Extend $\{C_j^1\}_{j=1}^J$ to a pants decomposition $\mathcal{P}$ of F, and regard $\mathcal{P}$ as a train track $\tau \subset F$ (choosing a bivalent switch on each component of $\mathcal{P}$). Of course, τ supports a measure

$$\mu_C(b) = \begin{cases} n_j, & \text{if } b \subset C_j^1 \text{ for some } j; \\ 0, & \text{otherwise,} \end{cases}$$

and the measured track (τ, μ_C) corresponds to G_C.

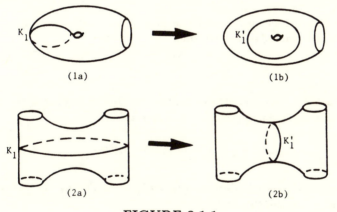

FIGURE 3.1.1

We build a generic birecurrent supertrack $\sigma \supset \tau$ by first defining σ in a neighborhood of $\mathcal{P}$ as in Figure 3.1.1a. The complement in F of this neighborhood consists of a disjoint union of surfaces of type $F_0^{s,r}$, where $r + s = 3$ and $r \geq 1$, and we define σ in each such region as indicated in Figure 3.1.1b-d. Arguing as in Proposition 2.6.1, one shows that σ is a birecurrent train track.

We define a measure $\mu_\emptyset$ on σ as follows: if b is a branch of σ contained in a pants curve in $\mathcal{P}$, then

$$\mu_\emptyset(b) = \begin{cases} 1, & \text{if } b \text{ is large;} \\ 3, & \text{if } b \text{ is small.} \end{cases}$$

This assignment extends uniquely to a positive measure $\mu_\emptyset$ on σ, as one checks without difficulty.

Now, for any subset $\mathcal{Q} \subset \mathcal{P}$, consider the function

$$\mu_\mathcal{Q}(b) = \begin{cases} 2, & \text{if } b \subset \mathcal{Q} \text{ is large;} \\ 0, & \text{if } b \subset \mathcal{Q} \text{ is small;} \\ \mu_\emptyset(b), & \text{if } b \not\subset \mathcal{Q}. \end{cases}$$

One easily checks that $\mu_\mathcal{Q}$ is a positive measure on σ for any $\mathcal{Q} \subset \mathcal{P}$, and we claim that for some $\mathcal{Q} \subset \mathcal{P}$, the multicurve $D_\mathcal{Q}$ corresponding to $\mu_\mathcal{Q}$ is connected. To see this, begin with the measure $\mu_\emptyset$. If $D_\emptyset$ is not connected, then there must be some pants curve $K \in \mathcal{P}$ which meets two distinct components of $D_\emptyset$. By construction, $D_{\{K\}}$ has one fewer component than $D_\emptyset$. Continuing in this way, we finally produce the putative connected multiple curve $D = D_\mathcal{Q}$ for some $\mathcal{Q} \subset \mathcal{P}$ and let $\mu = \mu_\mathcal{Q}$ denote the corresponding measure on σ. Notice that D meets each component of $\mathcal{P}$ exactly twice by construction.

The measure μ_C above on $\tau \subset \sigma$ gives rise to a measure (also denoted μ_C) on σ. For each $\ell \in \mathbb{Z}_+$, consider the measure

$$\mu_\ell = \mu + (2\ell)\,\mu_C$$

on σ. Since D is connected, the multicurve D_ℓ corresponding to (σ, μ_ℓ) is connected as well. Furthermore, by construction,

$$\mu_C = \lim_{\ell \to \infty} (2\ell)^{-1} \mu_\ell,$$

so, as above, G_C is the limit of $(2\ell)^{-1} G_{D_\ell}$, completing the proof of our claim and hence the theorem.   q.e.d.

One may thus think of $\mathcal{PL}_0(F)$ as a manifold (in fact, spherical) completion of the discrete set $\mathcal{S}'(F)$.

Actually, the manifold structure of $\mathcal{ML}_0(F)$ and $\mathcal{PL}_0(F)$ is much richer than that of topological manifolds, as we next discuss. We say that a piecewise-linear manifold is *piecewise-integral-linear* (or PIL for short) with respect to a choice of charts if the transition functions are piecewise-linear maps with integral coefficients. Similarly, a piecewise-projective manifold is *piecewise-integral-projective* (or PIP for short) if the transition functions are projective piecewise-linear with integral coefficients. Having made these definitions, we can state

Theorem 3.1.4: *For any surface $F = F_g^s$, the charts corresponding to complete train tracks in F give $\mathcal{ML}_0(F)$ the natural structure of an open PIL ball of dimension $6g - 6 + 2s$ and $\mathcal{PL}_0(F)$ the natural structure of a PIP sphere of dimension $6g - 7 + 2s$.*

Proof: First observe that the standard train tracks give $\mathcal{K}(F)$ a piecewise-linear structure by definition. Insofar as $\mathcal{ML}_0(F)$ is homeomorphic to $\mathcal{K}(F)$ by Corollary 2.8.6, $\mathcal{ML}_0(F)$ also has the structure of a piecewise-linear manifold. On the other hand, this structure seems to depend upon our choice of standard train tracks in F, which, in turn, depends upon our choice of standard basis for multicurves in F. To see that the piecewise-linear structure on $\mathcal{ML}_0(F)$ is natural, we must check that this structure is independent of these choices.

To this end, suppose first that F has no punctures, $\mathcal{P}$ is a pants decomposition of F, and $K \in \mathcal{P}$ is a pants curve. The component R of $(F - \mathcal{P}) \cup \{K\}$ containing K is either a surface of type $F_1^{0,1}$ or a surface of type $F_0^{0,4}$, and we may identify R with the appropriate surface in such a way that K coincides with the curve K_1 indicated in Figure 3.1.2(1a) or 3.1.2(2a), respectively. We then alter $\mathcal{P}$ in R by replacing K_1 with the curve K_1' indicated in Figure 3.1.2(1b) or 3.1.2(2b), respectively, to produce a new pants decomposition of F; we say that the new pants decomposition arises from $\mathcal{P}$ by applying an *elementary move*. A basic fact [HT] due to Hatcher-Thurston is that finite compositions of elementary moves act transitively on the set of all pants decompositions of F.

In case F has punctures, the component of $(F - \mathcal{P}) \cup \{K\}$ containing K is either a surface of type $F_1^{s,r}$, where $s + r = 1$, or a surface of type $F_0^{s,r}$, where $s + r = 4$, and there are obvious analogues of the elementary moves so that finite compositions again act transitively on the collection of pants decompositions of F.

We may extend the pants decompositions in Figure 3.1.2 to bases for multiarcs on the corresponding surfaces and ask for the associated changes of basic parameters for multiarcs from one basis to the other in each case. The explicit formulas (which are the main result of [P1]) are given in the Addendum, and, by inspection, these coordinate changes are found to be piecewise-linear with integral coefficients. It follows that the piecewise-linear structure on $\mathcal{ML}_0(F)$ induced by standard train tracks is indeed natural.

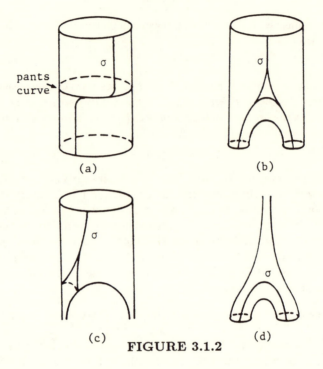

pants
curve

(a)   (b)

(c)   **FIGURE 3.1.2**   (d)

Now, given a positively measured complete train track (σ, λ) in F, we claim that there is a pants decomposition hitting σ efficiently. Indeed, arguing as in Theorem 3.1.3, one finds that pants decompositions are dense in $\mathcal{PL}_0(F)$. Furthermore, since σ is complete generic, it supports a positive tangential measure satisfying strict inequalities (in the definition of tangential measure) by Theorem 1.4.3. A nearby tangential measure therefore corresponds to the desired pants decomposition.

We may extend this pants decomposition to a basis $\mathcal{A}$ for multicurves in F. Collapsing (σ, λ) as in §2.7, we may assume that σ meets each pants curve transversely in a single point. Continuing to argue as in §2.7, we find that (σ, λ) refines to a track which is carried by a standard track for the basis $\mathcal{A}$. Consider the first split (if any) in this refinement which is a collision, say from σ' to σ'_O, and consider also the train tracks σ'_R and σ'_L which arise from a corresponding right and left split, respectively. According to Proposition 2.2.2

$$V(\sigma') = V(\sigma'_R) \cup V(\sigma'_O) \cup V(\sigma'_L),$$

so $V(\sigma'_R) \cup V(\sigma'_O) \cup V(\sigma'_L)$ contains a neighborhood of $[\sigma, \lambda]$ in $V(\sigma')$. It follows easily that there is a finite collection of complete train tracks $\sigma'_i < \sigma$, for $i = 1, \ldots, n$, so that $\cup_{i=1}^n V(\sigma'_i)$ covers a neighborhood of $[\sigma, \lambda]$, and each

σ_i' is carried by a train track τ_i which is standard for the basis $\mathcal{A}$. Thus, the transformation from $\cup_{i=1}^n V(\sigma_i')$ to the basic parameters with respect to the basis $\mathcal{A}$ is given by a piecewise-linear map.

By construction, τ_i refines to σ_i', and there can be no collisions in this refinement since σ_i' is complete, for $i = 1, \ldots, n$. Explicit formulas for splitting and shifting were given in §2.1, and incidence matrices associated with splitting and shifting were found to be unimodular (cf. the exercise at the end of §2.1) provided no split is a collision. The linear transformations comprising the piecewise-linear map from $\cup_{i=1}^n V(\sigma_i')$ to the basic parameters with respect to $\mathcal{A}$ are therefore invertible.

It follows from these remarks that if σ_1 and σ_2 are complete train tracks in F with

$$\mathcal{U} = \overset{\circ}{V}(\sigma_1) \cap \overset{\circ}{V}(\sigma_2) \neq \emptyset,$$

then the transition function on $\mathcal{U}$ associated with the change of chart from $\overset{\circ}{V}(\sigma_1)$ to $\overset{\circ}{V}(\sigma_2)$ is piecewise-linear with rational coefficients. To see that the coefficients are actually integral, recall that the collection of integral measures on a train track $\tau \subset F$ corresponds to the set of multicurves in F. Since the transition functions must preserve this locus, the coefficients must be integral, so charts associated to complete train tracks actually induce a PIL structure on $\mathcal{ML}_0(F)$. This PIL structure evidently induces a PIP structure on $\mathcal{PL}_0(F)$, and the proof is complete.   q.e.d.

Remark: Thus, the transition functions regarded as maps on the vector spaces of weights on branches are Lebesgue-measure-preserving transformations of $\mathbb{R}^{18g-18+6s}$, so $\mathcal{ML}_0(F)$ supports a natural class of measures.

§3.2 THE SYMPLECTIC STRUCTURE OF $\mathcal{ML}_0$

If $\tau \subset F$ is a generic birecurrent train track, then we let $W(\tau)$ denote the vector space of all assignments of (not necessarily nonnegative) real numbers, one to each branch of τ, which satisfy the switch conditions. In case τ is complete, we have seen that $\overset{o}{V}(\tau)$ can be thought of as a chart on the PIL manifold $\mathcal{ML}_0(F)$. As such, we may identify $W(\tau)$ with the tangent space to $\mathcal{ML}_0(F)$ at a point in $\overset{o}{V}(\tau)$. If $\sigma < \tau$, then there is an induced inclusion $W(\sigma) \subset W(\tau)$, which is given in coordinates by the same linear map which describes the inclusion $V(\sigma) \subset V(\tau)$. Thus, a symplectic structure on the PIL manifold $\mathcal{ML}_0(F)$ is given by a family

$$\{\cdot, \cdot\}_\tau : W(\tau) \times W(\tau) \to \mathbb{R}$$

of skew-symmetric, non-degenerate, bilinear pairings, where there is one such pairing for each complete train track $\tau \subset F$. Of course, we must check that the corresponding symplectic structure is independent of the choice of complete train track used as chart. We shall find that homology intersection numbers of real cycles in F induce such a family and the values of the symplectic pairings are invariant under splitting and shifting of train tracks.

Let τ be a train track in F, let v be a switch of τ with incident half-branches $\hat{a}$, $\hat{b}$, $\hat{c}$, where $\hat{a}$ is large, and let a, b, and c, respectively, denote the corresponding branches of τ. Suppose, moreover, that F is oriented so that $\hat{b}$ lies to the left and $\hat{c}$ lies to the right of v as we traverse $\hat{a}$ towards v; see Figure 3.2.1. Given tangent vectors $\xi, \eta \in W(\tau)$, we define the *symplectic pairing*

$$\{\xi, \eta\}_\tau = \frac{1}{2} \sum \begin{pmatrix} \xi(c) & \xi(b) \\ \eta(c) & \eta(b) \end{pmatrix},$$

where the sum is over all switches v of τ.

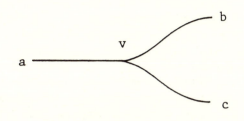

FIGURE 3.2.1

Lemma 3.2.1: *For any train track τ, $\{\cdot,\cdot\}_\tau$ is a skew-symmetric bilinear pairing on $W(\tau)$. Furthermore, if τ refines to σ (so $V(\sigma) \subset V(\tau)$), then*

$$\{\cdot,\cdot\}_\sigma = \{\cdot,\cdot\}_\tau|_{W(\sigma)}.$$

Proof: The first assertion is obvious. The second assertion follows easily by checking the various possibilities of refinement and is left as an exercise. q.e.d.

Thus, $\cup_\tau(\{\cdot,\cdot\}_\tau)$ defines a skew-symmetric bilinear pairing on the tangent space to $\mathcal{ML}_0(F)$. In order to see that the pairing is non-degenerate, we must introduce another formulation of these pairings, as follows.

Recall (from §1.3) that a train track $\tau \subset F$ is said to be orientable if there is an orientation of its branches that is consistent along incident edges, and that if a train track is orientable, then the frontier edge of each complementary region is composed of an even number of branches (where branches are counted with multiplicity). If τ is oriented, then there is a map

$$h_\tau : W(\tau) \longrightarrow H_1(F;\mathbb{R}),$$

where $\xi \in W(\tau)$ gives rise to a corresponding real cycle in the natural way. Namely, choose a CW decomposition of F which contains the CW complex underlying τ as a subcomplex, and take the value of ξ on a given branch as the coefficient of the corresponding cell, where the cell is given the orientation determined by the orientation of τ. This specification is evidently independent of the choice of CW decomposition of F, so the map h_τ above is well-defined.

Lemma 3.2.2: *Suppose that τ is a connected, orientable, recurrent train track in F and choose an orientation on τ. If $\xi, \eta \in W(\tau)$, then $\{\xi,\eta\}_\tau$ is the homology intersection number of the classes $h_\tau(\xi)$ and $h_\tau(\eta)$.*

Proof: Let N be a neighborhood of τ in F as in Step 1 of Construction 1.7.7 (for some arbitrary choice of positive measure on τ). Consider two isotopic copies τ_1 and τ_2 of τ in N. Near a switch v of τ, suppose that τ_1 and τ_2 are situated as illustrated in Figure 3.2.2, and adopt the notation indicated there for the branches of τ near v. The isotopy of τ to τ_i induces a natural identification $W(\tau) = W(\tau_i)$, for $i = 1, 2$. Of course, the homology intersection number of $h_\tau(\xi)$ and $h_\tau(\eta)$ agrees with the homology intersection number of $h_{\tau_1}(\xi)$ and $h_{\tau_2}(\eta)$, and this is computed easily from Figure 3.2.2 to be $\frac{1}{2}\{\xi(c)\eta(b) - \xi(b)\eta(c)\}$, as was asserted. We remark parenthetically that a different choice of orientation on τ does not affect the argument since homology intersection is itself bilinear.   q.e.d.

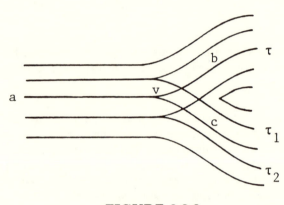

FIGURE 3.2.2

Now, let $\tau \subset F$ be an arbitrary (not necessarily orientable) train track, and consider the two-fold branched cover $\pi : \tilde{F} \to F$ with one branch point in each complementary region of τ in F. The full pre-image $\tilde{\tau} \subset \tilde{F}$ of τ is topologically the unbranched orientation cover of τ (in the sense of graph theory) and is therefore in fact an orientable train track. An example of such a covering $\pi : (\tilde{F}, \tilde{\tau}) \to (F, \tau)$ is given in Figure 3.2.3.

Let $I : \tilde{F} \to \tilde{F}$ denote the covering involution of π with induced homomorphism I_* acting $H_1(\tilde{F}; \mathbb{R})$ and let ω be a fixed choice of orientation on $\tilde{\tau}$. Thus, we have $I_*(\omega) = -\omega$. Moreover, if $\sigma \subset \tau$ is an orientable subtrack, then $\pi^{-1}(\sigma)$ consists of two components $\tilde{\sigma}_1$ and $\tilde{\sigma}_2$, and $I(\sigma_i) = \sigma_j$, for $\{i, j\} = \{1, 2\}$. On the other hand, if $\sigma \subset \tau$ is non-orientable, then $\pi^{-1}(\sigma)$ is connected.

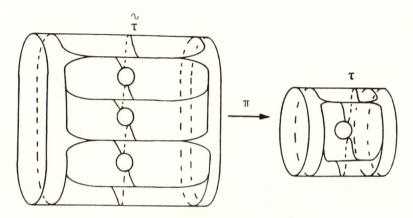

FIGURE 3.2.3

By Corollary 1.4.2, any train track is a subtrack of a complete train track, and we may assume in what follows that τ itself is complete. For convenience, we also assume for now that $F \neq F_1^1$, so that each complementary region of τ in F is either a trigon or a once-punctured monogon. Thus, each complementary region of $\tilde{\tau}$ in $\tilde{F}$ is either a hexagon or a once-punctured bigon, and one computes that the Euler characteristic of $\tilde{F}$ is four times the Euler characteristic of F using the Gauss-Bonnet and Riemann-Hurwitz formulas. Moreover, each puncture of $F = F_g^s$ has two distinct lifts to $\tilde{F}$, and it follows that the real dimension of $H_1(\tilde{F}; \mathbb{R})$ is

$$\dim_{\mathbb{R}} H_1(\tilde{F}; \mathbb{R}) = \begin{cases} 8g - 6, & \text{if } s = 0; \\ 8g - 7 + 4s, & \text{if } s \neq 0. \end{cases}$$

A standard spectral sequence argument (see, for instance, [Sp]) shows that the $+1$ eigenspace E^+ of I_* on $H_1(\tilde{F}; \mathbb{R})$ is isomorphic to $H_1(F; \mathbb{R})$, so

$$\dim_{\mathbb{R}} E^+ = \begin{cases} 2g, & \text{if } s = 0; \\ 2g + 2s - 1, & \text{if } s \neq 0. \end{cases}$$

Thus, the -1 eigenspace E^- of I_* in $H_1(\tilde{F}; \mathbb{R})$ has dimension $6g - 6 + 2s$. For the surface $F = F_1^1$, one similarly finds that the -1 eigenspace has real dimension two.

The map h_τ discussed above gives rise to a map

$$k_\tau : W(\tau) \rightarrow H_1(\tilde{F}; \mathbb{R})$$
$$\xi \mapsto h_{\tilde{\tau}}(\tilde{\xi}),$$

where $\tilde{\xi} \in W(\tilde{\tau})$ satisfies $\tilde{\xi} = \xi \circ \pi$.

Lemma 3.2.3: k_τ *is an isomorphism of* $W(\tau)$ *onto* $E^- \subset H_1(\tilde{F}; \mathbb{R})$.

Proof: The image of k_τ is clearly contained in E^-, and since

$$\dim_{\mathbb{R}} W(\tau) = 6g - 6 + 2s = \dim_{\mathbb{R}} E^-,$$

it suffices to show that k_τ is monic on $W(\tau)$. To this end, if τ has branches $\{b_i\}_1^n$, then choose respective lifts $\{\tilde{b}_i^1\}_1^n$ to $\tilde{\tau}$, and let $\tilde{b}_i^2 = I(\tilde{b}_i^1)$, for $i = 1, \ldots, n$. Consider the CW decomposition of $\tilde{F}$ with $\tilde{\tau}$ as one-skeleton so that $\{\tilde{b}_i^1, \tilde{b}_i^2\}_1^n$ forms a basis for one-chains in $\tilde{F}$, where each branch is taken with its orientation determined by ω. We may take the complementary regions of $\tilde{\tau}$ in $\tilde{F}$ (oriented arbitrarily) as a basis for the two-chains (with non-compact support) in $\tilde{F}$. It follows that I acts cellularly on our CW decomposition of $\tilde{F}$, and we let ∂ denote the boundary map on two-chains of this complex.

Now, if R is a component of $\tilde{F} - \tilde{\tau}$ with

$$\partial R = \sum_{i=1}^{n} \left(u_i^1 \, \tilde{b}_i^1 + u_i^2 \, \tilde{b}_i^2 \right)$$

for some $u_i^j \in \mathbb{R}$, for $i = 1, \ldots, n$ and $j = 1, 2$, then one sees immediately that $u_i^1 = -u_i^2$. Moreover, if $\xi \in W(\tau)$ and $k_\tau(\xi)$ is represented by the one-chain

$$\sum_{i+1}^{n} \left(v_i^1 \, \tilde{b}_i^1 + v_i^2 \, \tilde{b}_i^2 \right)$$

for some $v_i^j \in \mathbb{R}$, for $i = 1, \ldots, n$ and $j = 1, 2$, then it follows from the definition of k_τ that $v_i^1 = v_i^2$. These remarks show that the image of k_τ is disjoint from the image of ∂, and the lemma follows.   q.e.d.

To conclude, we have

Theorem 3.2.4: *For any surface* F, *the symplectic pairing is a skew-symmetric, non-degenerate, bilinear pairing on the tangent space to the PIL manifold* $\mathcal{ML}_0(F)$.

Proof: We first consider the case in which $F \neq F_1^1$. In light of Lemma 3.2.1, it remains to show that the pairing is non-degenerate. The previous two lemmas show that $\{\cdot, \cdot\}_\tau$ may be computed for any train track $\tau \subset F$ by passing to an oriented cover $\tilde{F}$, computing homology intersection numbers in $\tilde{F}$, and dividing by two. (This last two comes from the two-fold cover, not from the $\frac{1}{2}$ in the definition of the symplectic pairing.) Since the homology intersection pairing on $\tilde{F}$ is natural with respect to the covering involution I of $\tilde{F}$ and is non-degenerate, its restriction to the -1 eigenspace E^- of I_* is also non-degenerate. Since the image of k_τ has interior in E^-, we conclude that $\{\cdot,\cdot\}_\tau$ is non-degenerate on $W(\tau)$, as desired.

Finally, any complete train track in $F = F_1^1$ is orientable, so the symplectic pairing is induced by homology intersection numbers on F itself by Lemma 3.2.2. Non-degeneracy of the symplectic pairing follows as before, and the proof is complete.   q.e.d.

Remark: In fact, provided F has at least one puncture, the symplectic pairing on $\mathcal{ML}_0(F)$ is actually the pull-back of a smooth global symplectic form on a vector space by a piecewise linear imbedding of $\mathcal{ML}_0(F)$ into this vector space; see [PP2].

§3.3 TOPOLOGICAL EQUIVALENCE

If λ_1 and λ_2 are measures on a geodesic lamination G of compact support, then we say that the measured geodesic laminations (G, λ_1) and (G, λ_2) are *topologically equivalent*, and we let $\Delta(G)$ denote the *topological equivalence class* of all measured geodesic laminations with underlying geodesic lamination G. In case G is carried by a birecurrent train track τ, then measures λ_1 and λ_2 on G give rise to measures μ_1 and μ_2, respectively, on τ, and it follows from Theorem 1.7.12 that $\mu_1 = \mu_2$ if and only if $\lambda_1 = \lambda_2$. A geodesic lamination G of compact support is said to be *uniquely ergodic* if there is a unique projective class of measures of full support on G, so if G is uniquely ergodic, then $\Delta(G)$ is a ray in $\mathcal{ML}_0(F)$. A simple example of a non-uniquely ergodic geodesic lamination is given by taking G to be the disjoint union of two simple closed geodesics; indeed, we may assign measures to the components of G independently, so $\Delta(G)$ is two-dimensional in this case. (It follows from [M] that almost every measured geodesic lamination of compact support is uniquely ergodic.)

Suppose that (G, λ) is a measured geodesic lamination carried by the generic birecurrent train track $\tau \subset F$ with positive measure $\mu > 0$. We investigate here the topological equivalence class $\Delta(G) \subset V(\tau)$ of G. Choose, once and for all, a linear ordering on the switches of τ. Our immediate goal is to define a new measured train track $\tau' \subset F$ plus a linear ordering on the switches of τ' together with a choice of "symbol" $\zeta \in \{R, O, L, B\}$.

To do this, let u denote the first switch of τ (in the linear order) with corresponding singular point s of the bifoliated neighborhood $(N, \mathcal{F}, \mathcal{F}^\perp)$ of τ associated to μ (as in Step 1 of Construction 1.7.7); let ς denote the singular leaf of $\mathcal{F}$ issuing forth from s. Consider the large one-way trainpath (cf. §2.3) ρ on τ starting from u and imagine unzipping (cf. §2.4) $(N, \mathcal{F}, \mathcal{F}^\perp)$ along the corresponding initial segment of ς. There are three cases as follows: it may be that ς is a finite singular leaf corresponding to ρ; if not, then ς travels either to the right or to the left of the final point v of a tie-transverse path (cf. the proof of Theorem 1.7.8) corresponding to ρ (in its natural orientation starting from s). In the former case, completely unzip along ς, and in the latter cases, unzip along ς slightly beyond (to preserve genericity) the singular tie containing v. Collapsing ties of the unzipped neighborhood produces a measured generic train track (τ', μ'), and τ' arises from τ by a (possibly empty) finite sequence of shifts, yielding a train track σ, followed by a single split, say along the edge e of σ.

Of course, we may split σ along e in three ways, and according to Lemma 2.1.3, either all three possibilities are recurrent or exactly one is recurrent. If all three are recurrent, then the symbol ζ is defined to be $R, O,$ or L depending on whether the split along e is a right split, a collision, or a left split, respectively. If only one of the three is recurrent, then the symbol ζ is defined to be B (for blank). Finally, the linear order on the switches of τ' is derived from the linear order on the switches of τ by omitting the switches of τ corresponding to the singular points u and v from the order in case ς is a finite singular leaf corresponding to ρ and then cyclically advancing by one the linear order (using the natural identification of switches of τ' with the remaining switches of τ).

We may iterate this process (until, perhaps, there are no switches) to produce a sequence

$$(\tau, \mu) = (\tau_0, \mu_0), (\tau', \mu') = (\tau_1, \mu_1), (\tau_2, \mu_2), \ldots$$

of measured train tracks and a sequence

$$\zeta = \zeta_0, \zeta_1, \ldots$$

of symbols. Of course, the sequence terminates if and only if (τ, μ) corresponds to a family of weighted multiple curves in F. Finally, omitting the symbol B from this sequence wherever it occurs, we produce the *splitting sequence* $\zeta_\tau(\mu)$ of μ, which is regarded as a (finite or semi-infinite) word in the alphabet $\{R, O, L\}$.

Theorem 3.3.1: *Suppose that τ is a birecurrent generic train track, and let $\mu_1, \mu_2 \in \overset{o}{V}(\tau)$. The measured laminations corresponding to (τ, μ_1) and (τ, μ_2) are topologically equivalent if and only if*

$$\zeta_\tau(\mu_1) = \zeta_\tau(\mu_2).$$

Moreover, if ζ is an initial segment of $\zeta_\tau(\mu)$ for some $\mu \in \overset{o}{V}(\tau)$, then either $\zeta = \zeta_\tau(\mu)$, or each ζ, X is an initial segment of $\zeta_\tau(\mu')$ for some $\mu' \in \overset{o}{V}(\tau)$, for each $X \in \{R, O, L\}$.

Proof: Suppose first that μ_1 and μ_2 correspond to topologically equivalent measured geodesic laminations, say with underlying geodesic lamination G. The splitting sequences $\zeta_\tau(\mu_1)$ and $\zeta_\tau(\mu_2)$ coincide by construction since the leaves of G determine the symbol at each stage.

Conversely, suppose that μ_1 and μ_2 have the same splitting sequence, and let $(N_i, \mathcal{F}_i, \mathcal{F}_i^\perp)$ denote the bifoliated neighborhood of τ corresponding to μ_i, for $i = 1, 2$. Evidently, the singular leaves of $\mathcal{F}_i$ are represented by the same trainpaths on τ, for $i = 1, 2$, and the topological equivalence of the corresponding measured geodesic laminations follows easily from Corollary 1.7.10.

The last assertion follows readily from Lemma 2.1.3, completing the proof.   q.e.d.

Remark: It is an easy matter to construct a word in the alphabet $\{R, O, L\}$ which does not arise as the splitting sequence of some measure. For instance, for the complete train track on the surface F_1^1, the word $LLL \ldots$ cannot occur.

To conclude, we give an explicit description of the closure $\bar{\Delta}(G)$ of $\Delta(G)$ in $V(\tau)$, as follows.

Proposition 3.3.2: *Given a birecurrent positively measured generic train track (τ, μ) corresponding to a measured geodesic lamination with underlying geodesic lamination G, let*

$$\tau = \tau_0 > \tau_1 > \ldots$$

be the sequence of birecurrent train tracks described in Corollary 1.7.13. Then

$$\bar{\Delta}(G) = \cap_{m \geq 0} V(\tau_m).$$

Proof: As above, the tracks τ_m are determined by the leaves of G, so $\Delta(G) \subset \cap_{m \geq 0} V(\tau_m)$. Insofar as $\cap_{m \geq 0} V(\tau_m)$ is closed in $V(\tau)$, one inclusion follows. Conversely, if $(G', \lambda') \in \cap_{m \geq 0} V(\tau_m)$, then $G' < \tau_m$ for each $m \geq 0$, so, by Theorem 1.6.6, $E(\tilde{G}') \subset \cap_{m \geq 0} E(\tau_m) = E(\tilde{G})$. Thus, $G' \subset G$, so $(G', \lambda') \in \bar{\Delta}(G)$, as desired.   q.e.d.

§3.4 DUALITY AND TANGENTIAL COORDINATES

A one-dimensional CW complex $\tau \subset F$ is called a *bigon track* in F if the smoothness and non-degeneracy Conditions (1) and (2) in the definition of train track (see §1.1) hold for τ as well as the condition

(3') No component of $F - \tau$ is an imbedded nullgon, monogon, once-punctured nullgon, or annulus.

Thus, we are simply relaxing Condition (3) in the definition of train track to allow complementary bigons.

A *transverse measure* μ on the bigon track τ is an assignment of a nonnegative real number $\mu(b)$ to each branch b of τ satisfying the switch conditions, as before, and the collection of all transverse measures on τ is denoted $V(\tau)$. We claim that an integral transverse measure on a bigon track determines a multiple curve (by concatenating arcs parallel to the branches, as before). To see this, mimic the proof of Proposition 1.1.1 (and apply this result as in §1.2), where each complementary bigon is foliated as illustrated in Figure 3.4.1. More generally, given a positive transverse measure on a bigon track $\tau \subset F$, mimic Step 1 of Construction 1.7.7 to produce a bifoliated neighborhood $(N, \mathcal{F}, \mathcal{F}^{\perp})$ of τ in F. One can furthermore mimic Step 2 of Construction 1.7.7 to produce a measured lamination in F. (The construction of a measured geodesic lamination from a transverse measure on a bigon track will be discussed below.) A bigon track is *generic* if each switch is trivalent (or bivalent and lies on some simple closed curve component of the bigon track). If $\mu \in V(\tau)$ for some generic bigon track $\tau \subset F$, then we may alter (τ, μ) by splitting or shifting, where the definitions are exactly as in §2.1.

191

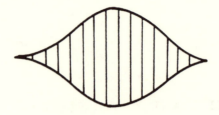

FIGURE 3.4.1

A *tangential measure* ν on τ is an assignment $\nu(b)$ to each branch b of τ satisfying Conditions (1) and (2) in the definition of tangential measure on train tracks (see §1.3) as well as the condition

> (3) The total tangential measures of the frontier edges of a complementary bigon must be equal.

The collection of all tangential measures on τ is denoted $V^*(\tau)$.

A bigon track is said to be *recurrent* if it supports a strictly positive transverse measure and is said to be *transversely recurrent* if it supports a strictly positive tangential measure. A bigon track which is both recurrent and transversely recurrent is said to be *birecurrent*. A bigon track in a surface other than F_1^1 is called *maximal* if each complementary region is a trigon, a once-punctured monogon, or a bigon; a bigon track in F_1^1 is called *maximal* if each complementary region is a (perhaps once-punctured) bigon. A maximal birecurrent bigon track is said to be *complete*.

A bigon track *carries* another bigon track (or a lamination) if there is a supporting map satisfying Conditions (1)-(3) in the definition of carrying by train tracks (see §1.2). If the bigon track τ carries the bigon track σ, then we write $\sigma < \tau$ or $\tau > \sigma$, and a transverse measure on σ gives rise to a transverse measure on τ as determined by an incidence matrix associated to a supporting map for the carrying (as in §2.1). If the bigon track τ carries the measured lamination (L, λ), then we write $L < \tau$ or $\tau > L$, and there is an induced transverse measure on τ (as in §1.7).

Two bigon tracks (or a bigon track and a lamination) in a surface F are said to *hit efficiently* if there is no imbedded bigon in F whose frontier is composed of two C^1 segments, one from each of the bigon tracks (or one from the bigon track and one contained in leaf of the lamination).

Given a bigon track $\tau \subset F$, enumerate the bigon components of $F - \tau$. We may iteratively collapse these bigon components in the chosen order (making further choices as well) onto arcs connecting the vertices in the frontiers of the bigons as indicated in Figure 3.4.2 to produce a train track

$\bar{\tau} \subset F$. Clearly, $\bar{\tau}$ depends on our choices, and $\bar{\tau}$ is a maximal train track if and only if τ is a maximal bigon track.

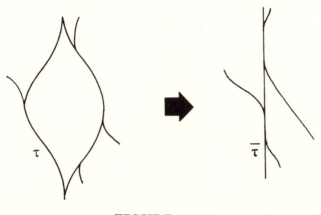

FIGURE 3.4.2

We claim that $\bar{\tau}$ is recurrent if τ is. Indeed, insofar as $\bar{\tau} > \tau$, there is an associated incidence matrix M, and since any supporting map for the carrying maps τ onto $\bar{\tau}$, M associates a positive measure on $\bar{\tau}$ to a positive measure on τ. In contrast, $\bar{\tau}$ may fail to be transversely recurrent even if τ is transversely recurrent. On the other hand, if $\mu \in V(\tau)$ gives rise to the measure $\bar{\mu} \in V(\bar{\tau})$, then we may apply the extension to Construction 1.7.7 (in Theorem 2.7.4) to $(\bar{\tau}, \bar{\mu})$ to produce a measured geodesic lamination in F. Evidently, the measured lamination associated to $(\bar{\tau}, \bar{\mu})$ in Step 2 of Construction 1.7.7 is isotopic to the measured lamination constructed above from (τ, μ) (in analogy to Step 2 of Construction 1.7.7), and the associated measured geodesic lamination is therefore independent of our choices. Denote the corresponding map by $\psi_\tau : V(\tau) \to \mathcal{ML}_0(F)$, and define the quotient

$$V'(\tau) = V(\tau)/\sim, \text{ where } \mu_1 \sim \mu_2 \text{ if } \psi_\tau(\mu_1) = \psi_\tau(\mu_2).$$

According to Theorem 2.7.4, if $\mu_1 \sim \mu_2$, then the corresponding transverse measures $\bar{\mu}_1, \bar{\mu}_2 \in V(\bar{\tau})$ must be equal, and the converse holds by definition. Of course, if τ is actually a train track, then $V'(\tau) = V(\tau)$.

It follows that $V'(\tau)$ is a linear quotient of the cone $V(\tau)$. Indeed, regarding each $\mu_i \in V(\tau)$ as a vector of weights assigned to the branches of τ, for $i = 1, 2$, $\mu_1 \sim \mu_2$ if and only if $\mu_1 - \mu_2$ lies in the kernal of an incidence matrix for the carrying $\tau < \bar{\tau}$ by Theorem 2.7.4. In particular, even though the incidence matrix depends on the choice of supporting map (see the remark following Theorem 2.7.4), the kernal is independent of this choice.

Summarizing, we have

Proposition 3.4.1 *If $\tau \subset F$ is a recurrent bigon track, then Construction 1.7.7 determines a continuous injection of $V'(\tau)$ into $\mathcal{ML}_0(F)$. Indeed, identifying $V'(\tau)$ with the corresponding subset of $\mathcal{ML}_0(F)$, we have*

$$V'(\tau) = \{(G, \lambda) \in \mathcal{ML}_0(F) : G < \tau\},$$

and the interior $\overset{\circ}{V}{}'(\tau)$ is open in $\mathcal{ML}_0(F)$ if τ is complete.

We say that a bigon track $\tau \subset F$ *fills* F if any multiple curve in F intersects τ. Thus, τ fills F if and only if each component of $F - \tau$ is a (perhaps once-punctured) topological disk.

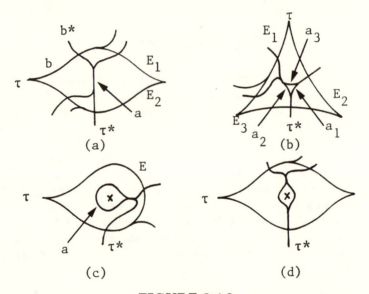

FIGURE 3.4.3

If τ is a bigon track (which may, in particular, be a train track) which fills F, then we define the *dual bigon track* $\tau^* \subset F$ as follows. For each branch b_i of τ, choose a short arc b_i^* meeting τ transversely in a single point of b_i so that the arcs b_i^* are pairwise disjoint. Suppose first that R is a simply connected complementary region of $F - \tau$, and let E be a frontier edge of R which is composed of the branches $b_1, \ldots, b_k$. The transverse arcs $b_1^*, \ldots, b_k^*$ are all made confluent in R near E (and are incident on a vertex of valence $k+1$ if $k > 1$) to form a branch, say e, of τ^*. Finally, if e_1 and e_2 arise in this way and correspond to adjacent fronter edges E_1 and E_2 of R, then add a branch to τ^* smoothly connecting e_1 and e_2 in R. In particular, if R is a bigon, then we add only one branch connecting e_1 to e_2 as in Figure 3.4.3a. Figure 3.4.3b illustrates an example when R is a trigon. A similar procedure is employed for a once-punctured complementary region R; see Figures 3.4.3c and 3.4.3d for examples when R is a once-punctured monogon and bigon, respectively. (The branches labeled a or a_i in these figures will be used later.)

This construction describes the non-generic dual bigon track τ^* associated to τ. Of course, we could comb τ^* (as in §1.4) to a generic bigon track (which is determined only up to shifting), but we shall not do this. Each transverse arc b_i^* is contained in a unique branch (of the same name) of τ^*, and the branch b_i^* of τ^* is said to be *dual* to the corresponding branch b_i of τ. Notice that there is one bigon B_v complementary to τ^* for each switch v of τ as in Figure 3.4.4, and the closure of each such bigon is imbedded in F. Furthermore, τ^* hits τ efficiently by construction.

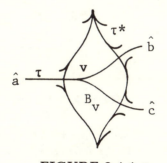

FIGURE 3.4.4

Our first glimpse of the duality between transverse and tangential measure is given by the following result, which will be elaborated upon in the Epilogue.

Proposition 3.4.2: *If $\tau \subset F$ is a generic bigon track filling F which refines (that is, splits and shifts) to the bigon track σ with no collisions (so $\sigma < \tau$ fills F as well), then $\tau^* < \sigma^*$. If there are furthermore no shifts in the refinement of τ to σ (so τ simply splits to σ), then there are supporting maps for the carryings $\sigma < \tau$ and $\tau^* < \sigma^*$ with respective incidence matrices M and its transpose M^t.*

Proof: First suppose that σ arises from τ by a single split, and consider Figure 3.4.5a, where we illustrate the four relevant bigon tracks. By inspection, we see that $\tau^* < \sigma^*$ and leave it as an exercise for the reader to construct supporting maps for the carryings $\sigma < \tau$ and $\tau^* < \sigma^*$ whose incidence matrices are transposes of one another.

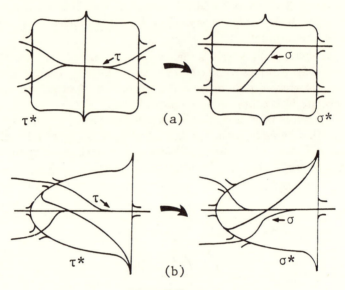

FIGURE 3.4.5

Next, suppose that σ arises from τ by a single shift and consider Figure 3.4.5b, where we illustrate the four relevant bigon tracks. Again, one sees by inspection that $\tau^* < \sigma^*$. The first assertion in the proposition follows directly from these observations and the transitivity of $<$.

Turning to the second assertion of the proposition, suppose that

$$\sigma = \tau_n < \tau_{n-1} < \ldots < \tau_0 = \tau$$

is a sequence of bigon tracks so that τ_i arises from τ_{i-1} by a single split, for $i = 1, \ldots, n$. According to the previous paragraph,

$$\tau^* = \tau_0^* < \tau_1^* < \ldots \tau_n^* = \sigma^*,$$

and there are supporting maps M_i and M_i^t for the carryings $\tau_i < \tau_{i-1}$ and $\tau_{i-1}^* < \tau_i^*$, respectively, for $i = 1, \ldots, n$. By functoriality of incidence matrices (cf. §2.1), we conclude that there are supporting maps for the carryings $\sigma < \tau$ and $\tau^* < \sigma^*$ with respective incidence matrices

$$M = M_1 \, M_2 \, \ldots \, M_n,$$
$$M^t = \left(M_1 \, M_2 \ldots M_n \right)^t = M_n^t \, \ldots M_2^t \, M_1^t,$$

and the proof is complete.   q.e.d.

If two bigon tracks τ and τ' in F hit efficiently, then there is a symmetric bilinear pairing

$$< \cdot, \cdot >_{\tau,\tau'} : V(\tau) \times V(\tau') \to [\mathbb{R}_+ \cup \{0\}]$$

defined as follows. Suppose that τ has branches $\{b_i\}_1^n$, τ' has branches $\{b_j'\}_1^{n'}$, and let m_{ij} denote the number of times b_i intersects b_j'. If $\mu \in V(\tau)$ and $\mu' \in V(\tau')$, then we define

$$< \mu, \mu' >_{\tau,\tau'} = \sum_{i=1}^{n} \sum_{j=1}^{n'} m_{ij} \ \mu(b_i) \ \mu'(b_j').$$

To motivate the definition, observe that if $\mu \in V(\tau)$ is integral and corresponds to the multicurve c in F, then

$$< \mu, \mu' >_{\tau,\tau'} = \mathcal{V}_{[\tau',\mu']}(c),$$

where $\mathcal{V}_{[\tau',\mu']}$ is the variation function of $[\tau', \mu']$ considered in the previous chapter.

Remark: Actually, the pairing above on $V(\tau) \times V(\tau')$ is the restriction of a non-degenerate, symmetric, locally bilinear pairing on $\mathcal{ML}_0(F)$ which continuously extends the geometric intersection function on $\mathcal{S}(F)$. See [Bo;§III] for a discussion of this "intersection pairing" on $\mathcal{ML}_0(F)$.

For convenience, we shall henceforth restrict attention to complete bigon tracks in a surface different from F_1^1 (and shall comment separately on the case of F_1^1 in subsequent remarks). Furthermore, some of the material we develop generalizes to the larger class of bigon tracks filling F, but there are some consequential differences. (We shall also mention some of this more general theory in subsequent remarks.)
        The basic statement of duality between tangential and transverse measure is given in our next result.

Lemma 3.4.3: *If $\tau \subset F \neq F_1^1$ is a complete bigon track, then there is a linear isomorphism*

$$V^*(\tau) \to V(\tau^*)$$

$$\nu \mapsto \nu^*$$

so that

$$< \mu, \nu^* >_{\tau,\tau^*} = \sum \mu(b) \ \nu(b),$$

for any $\mu \in V(\tau)$, where the sum is over all branches b of τ. Furthermore, ν^* is positive if and only if ν is positive and the total ν-tangential measures of frontier edges of each trigon complementary to τ in F satisfy the three strict triangle inequalities.

Proof: Suppose that $\nu \in V^*(\tau)$ is a tangential measure on τ, and first consider a trigon R complementary to τ in F. We claim that there is a unique transverse measure $\nu^* \in V(\tau^*)$ with the property that $\nu^*(b^*) = \nu(b)$ for each branch b of τ with dual branch b^* of τ^*. Indeed, adopting the notation of Figure 3.4.3b for the branches of τ^* inside R and the frontier edges of R, such a transverse measure ν^* must satisfy

$$\nu^*(a_i) + \nu^*(a_j) = \nu(E_k), \text{ for } \{i,j,k\} = \{1,2,3\},$$

and this system of equations admits the unique solution

$$\nu^*(a_i) = \frac{1}{2}\{\nu(E_j) + \nu(E_k) - \nu(E_i)\}, \text{ for } \{i,j,k\} = \{1,2,3\}.$$

Since ν is a tangential measure, the solution ν^* takes nonnegative values on a_1, a_2, and a_3. Furthermore, ν^* takes strictly positive values on these branches if and only if the total ν-tangential measures of E_1, E_2, and E_3 satisfy the strict triangle inequalities.

Similarly, adopting the notation of Figure 3.4.3c for the branches of τ^* inside a complementary once-punctured monogon and its frontier edges, one finds

$$\nu^*(a) \quad = \quad \frac{1}{2} \nu(E).$$

Finally, adopting the notation of Figure 3.4.3a for the branches of τ^* in a complementary bigon and its frontier edges, one finds

$$\nu^*(a) = \nu(E_1) = \nu(E_2).$$

In any case, the determined values of ν^* extend uniquely to a transverse measure ν^* on τ^* which is positive exactly under the stated conditions. The expression for the pairing follows from the definitions and previous remarks, completing the proof.   q.e.d.

Remark: A tangential measure on a complete bigon track in F_1^1 determines only a one-parameter family of transverse measures on its dual bigon track. More generally, in case $\tau \subset F$ only fills F, the result is actually false. Indeed, there are further restrictions on a tangential measure $\nu \in V^*(\tau)$ which must be satisfied in order that there is a *nonnegative* solution ν^* to the equations $\nu^*(b^*) = \nu(b)$. Moreover, even when these further conditions are satisfied, the solution is not unique: it turns out that if τ has k complementary (perhaps once-punctured) polygons with an even number of sides, then there is a k parameter family of transverse measures on τ^* solving these equations. The reader can consult [P2] for a further discussion of this.

Lemma 3.4.4: *Suppose that $\tau \subset F \neq F_1^1$ is a complete bigon track and $\sigma \subset F$ is a bigon track.*

> *(a) σ (is isotopic to a track which) hits τ efficiently if and only if $\sigma < \tau*$.*

> *(b) σ (is isotopic to a track which) hits τ^* efficiently if and only if $\sigma < \tau$.*

> *(c) The operation $\cdot^*$ is weakly reflexive in the sense that $\tau < \tau^{**} < \tau$ and $V'(\tau) = V'(\tau^{**}) \subset \mathcal{ML}_0(F)$.*

Proof: If $\sigma < \tau^*$, then a bigon in F with its two frontier arcs contained in σ and τ, respectively, gives rise to a bigon with its two frontier arcs contained in τ^* and τ, respectively. This contradicts that τ^* hits τ efficiently, and we conclude that σ hits τ efficiently if $\sigma < \tau^*$. Conversely, suppose that σ hits τ efficiently. Since τ is complete hence maximal, τ^* is easily seen to have the following property: if R is a component of $F - \tau$ and ρ is a trainpath on σ, then $\rho \cap R$ is smoothly and properly homotopic into $\tau^* \cap R$. Using this fact, it is easy to isotope σ into an appropriate neighborhood of τ^*. Thus, $\sigma < \tau^*$, completing the proof of part (a).

The proof of part (b) is similar, and together parts (a) and (b) show that $\tau < \tau^{**} < \tau$ since τ^* hits both τ and τ^{**} efficiently. The last assertion of part (c) follows from the first assertion of part (c) and the characterization of $V'(\tau)$ given in Proposition 3.4.1 above.   q.e.d.

Proposition 3.4.5: *If $\tau \subset F \neq F_1^1$ is a complete generic train track, then τ^* is a complete bigon track.*

Proof: Since τ is generic and transversely recurrent, it supports a positive tangential measure ν satisfying strict inequalities on complementary trigons by Theorem 1.4.3. By the last part of Lemma 3.4.3, the dual transverse measure ν^* on τ^* is positive, and we conclude that τ^* is recurrent.

Since τ is recurrent, it supports a positive integral transverse measure giving rise to a multiple curve C in F which we regard as lying in a small tie neighborhood of τ. The components of C hit τ^* efficiently by Lemma 3.4.4b and meet each branch of τ^* dual to a branch of τ. Inspection of Figure 3.4.3b and 3.4.3c shows that the analogue of sneaking up for non-generic tracks (discussed at the beginning of §1.4) can be employed to isotope C to produce curves hitting τ^* efficiently and meeting the remaining branches of τ^*. One uses these curves as in §1.3 to produce the required positive tangential measure on τ^*, and we conclude that τ^* is transversely recurrent as well.

Thus, τ^* is birecurrent. We remarked above that maximality of τ implies maximality of τ^*, and the proof is complete.   q.e.d.

Now, given a complete bigon track $\tau \subset F$, define the quotient

$$V^{*\prime}(\tau) = V^*(\tau)/ \sim, \text{ where } \nu_1 \sim \nu_2 \text{ if } \psi_{\tau^*}(\nu_1^*) = \psi_{\tau^*}(\nu_2^*).$$

Thus, the extension of Construction 1.7.7 determines a mapping

$$V^{*\prime}(\tau) \rightarrow \mathcal{ML}_0(F)$$
$$\nu \mapsto \psi_{\tau^*}(\nu^*),$$

and we find the following dual of Proposition 3.4.1.

Proposition 3.4.6: *If $\tau \subset F$ is a complete generic train track, then the extension of Construction 1.7.7 determines a continuous injection of $V^{*\prime}(\tau)$ into $\mathcal{ML}_0(F)$. Indeed, identifying $V^{*\prime}(\tau)$ with the corresponding subset of $\mathcal{ML}_0(F)$, we have*

$$V^{*\prime}(\tau) = \{(G,\lambda) \in \mathcal{ML}_0(F) : G \text{ hits a track isotopic to } \tau \text{ efficiently}\},$$

and the interior $\overset{\circ}{V}{}^{*\prime}(\tau)$ is open in $\mathcal{ML}_0(F)$.

Proof: According to Lemma 3.4.3, $V^*(\tau)$ is linearly isomorphic to $V(\tau^*)$, so $V^{*\prime}(\tau)$ is isomorphic to $V'(\tau^*)$. The first assertion therefore follows from the first assertion of Proposition 3.4.1. The second assertion of Proposition 3.4.1 identifies the subset of $\mathcal{ML}_0(F)$ corresponding to $V'(\tau^*)$, and the second assertion above then follows easily from Lemma 3.4.3a. Finally, since τ^* is complete if τ is by Proposition 3.4.5, the final assertion follows from the final assertion of Proposition 3.4.1.   q.e.d.

We wish to elucidate the equivalence relation $\sim$ on $V^*(\tau)$. To this end, suppose that v is a switch of the complete generic train track $\tau \subset F$, and let $\hat{a}, \hat{b}$, and $\hat{c}$ be the incident half-branches contained in respective branches a, b and c of τ, where $\hat{a}$ is large; see Figure 3.4.4. Suppose that $\nu_1, \nu_2 \in V^*(\tau)$ satisfy the condition

$$\{\nu_2(a) - \nu_1(a)\} = \{\nu_1(b) - \nu_2(b)\} = \{\nu_1(c) - \nu_2(c)\},$$

which is called the *bigon condition of v*. We say simply that two tangential measures on τ satisfy the *bigon conditions* if they satisfy the bigon condition of each switch of τ. The corresponding transverse measures ν_1^* and ν_2^* on τ^* give rise to the same measured geodesic lamination $\psi_{\tau^*}(\nu_1^*) = \psi_{\tau^*}(\nu_2^*)$. Indeed, the bigon condition of v corresponds to pulling leaves of the corresponding laminations (built as in Step 2 of Construction 1.7.7) across the bigon B_v complementary to τ^* associated to v.

If τ has n branches, regard

$$V^*(\tau) \subset \left[\mathbb{R}_+ \cup \{0\} \right]^n \subset \mathbb{R}^n,$$

by listing the values of a tangential measure on the branches of τ as an n-tuple. Similarly, regard

$$V(\tau) \subset \left[\mathbb{R}_+ \cup \{0\} \right]^n \subset \mathbb{R}^n$$

by listing (in the same order) the values taken by a transverse measure on the branches of τ. Define the equivalence relation

$$\mathcal{B} = \{ (\nu_1, \nu_2) \in V^*(\tau) \times V^*(\tau) : \nu_1, \nu_2 \text{ satisfy the bigon conditions} \}.$$

Collapse bigons complementary to τ^* as before (making choices) to produce a train track $\bar{\tau}^* \subset F$, let M be the natural incidence matrix for the carrying $\tau^* < \bar{\tau}^*$, and define the equivalence relation

$$\mathcal{M} = \{ (\mu_1, \mu_2) \in V(\tau^*) \times V(\tau^*) : M\mu_1 = M\mu_2 \}.$$

We claim that $\mathcal{B}$ is dual to $\mathcal{M}$ in the sense that for any pair ν_1, ν_2 of tangential measures on τ,

$$(\nu_1, \nu_2) \in \mathcal{B} \text{ if and only if } (\nu_1^*, \nu_2^*) \in \mathcal{M}.$$

Suppose first that $(\nu_1, \nu_2) \in \mathcal{B}$. In light of the remarks above, we conclude that $\psi_{\tau^*}(\nu_1^*) = \psi_{\tau^*}(\nu_2^*)$, so $(\nu_1^*, \nu_2^*) \in \mathcal{M}$ by the discussion before Proposition 3.4.1.

Conversely, suppose that $(\nu_1^*, \nu_2^*) \in \mathcal{M}$. Insofar as τ^* is recurrent, transverse measures on τ^* taking rational values are dense in $V(\tau^*)$, so it suffices to consider the case in which ν_1^* and ν_2^* are integral. Thus, we must show that if a multicurve in F is carried by τ^* in two different ways, then the corresponding supporting maps are related by pulling arcs across bigons complementary to τ^*, as above. To this end, construct the bifoliated neighborhood $(N, \mathcal{F}, \mathcal{F}^\perp)$ of τ^* corresponding to some positive transverse measure on τ^* (using Proposition 3.4.5), and suppose that $C_1, C_2 \subset N$ are multiple curves transverse to $\mathcal{F}^\perp$ which represent the same multicurve in F.

The proof is by induction on the number of times C_1 intersects C_2. If these multiple curves are disjoint, then consider the annulus A in F having them as boundary. Extend $\mathcal{F}^\perp$ to a foliation $\hat{\mathcal{F}}^\perp$ of F by foliating each complementary trigon and once-punctured monogon as in Figure 1.1.2 and each complementary bigon as in Figure 3.4.1. $\hat{\mathcal{F}}^\perp$ restricts to a foliation of A, and, by the Poincaré-Hopf Theorem, we conclude that A is composed of some collection of bigons. By construction, each such bigon is either contained entirely in N or contains some bigon B_v associated to a switch v of τ. Since the closure of each B_v is imbedded in F, one sees easily that C_1 and C_2 are related by pulling arcs across bigons, as desired.

For the inductive step, if $C_1 \cap C_2 \neq \emptyset$, then there must be an imbedded bigon B complementary to $C_1 \cup C_2$. If $B \subset N$, then an isotopy along $\mathcal{F}^\perp$ decreases the number of intersection points. If $B \not\subset N$, then $\hat{\mathcal{F}}^\perp$ restricts to a foliation of B, and the Poincaré-Hopf Theorem is applied as before.

Thus, the supporting maps of C_1 and C_2 are related by pulling arcs across bigons complementary to τ^*, as was asserted. It follows that if $(\nu_1^*, \nu_2^*) \in \mathcal{M}$, then $(\nu_1, \nu_2) \in \mathcal{B}$, and the proof of our claim is complete.

We summarize with

Theorem 3.4.7: *Suppose that $\tau \subset F \neq F_1^1$ is a complete generic train track. Then $V^{*\prime}(\tau)$ is the linear quotient of $V^*(\tau)$ by the equivalence relation $\mathcal{B}$ determined by the bigon conditions, and the linear isomorphism described in Lemma 3.4.3 descends to an identification*

$$V^{*\prime}(\tau) = V'(\tau^*)$$

of subsets of $\mathcal{ML}_0(F)$. Furthermore, $V^{*\prime}(\tau) \cap V(\tau)$ consists of the empty lamination.

Proof: Only the last sentence requires comment, and this follows easily from Propositions 3.4.1 and 3.4.6 since only the empty lamination is carried by a bigon track which it hits efficiently.   q.e.d.

Remark: For the surface F_1^1, we have observed in the remark following Lemma 3.4.3 that a tangential measure determines only a line in $\mathcal{ML}_0(F)$. One can projectivize to derive tangential coordinates on $\mathcal{ML}_0(F_1^1)$.

Thus, $\overset{\circ}{V}{}^{*\prime}(\tau)$ describes a coordinate patch on $\mathcal{ML}_0(F)$ if $\tau \subset F \neq F_1^1$ is a complete generic train track, and we have a notion of *tangential coordinates* on τ arising as the equivalence classes of tangential measures under the quotient by the bigon conditions.

One expression of the duality we have discussed is as follows. Suppose that $\tau \subset F$ is a (not necessarily recurrent or transversely recurrent) maximal train track. We saw in §1.3 that τ is recurrent if and only if $V(\tau)$ has interior in $\mathcal{ML}_0(F)$. Dually, we have found here that τ is transversely recurrent if and only in $V^{*\prime}(\tau)$ has interior in $\mathcal{ML}_0(F)$.

Furthermore, we remark that whereas a transverse measure on a train track can be thought of as an assignment of widths to the rectangles that comprise the corresponding bifoliated neighborhood of τ in F (as in Construction 1.7.7), a tangential measure on the train track can be thought of as the lengths of the rectangles comprising N. This will be elaborated upon in the Epilogue.

EPILOGUE

The purpose of this section is to briefly mention the larger contexts of Riemann surfaces and surface diffeomorphisms in which the considerations of this volume play a role. The material of this volume is also germaine to dynamical systems ([M] and [FLP] are good starting places) and to the geometry and topology of three-manifolds (see [T2] and [Bo]), but we do not attempt to discuss this material here. Our intent is simply to survey a portion of the relevant material mostly due to Thurston. We mention that [C] contains an introductory account of some of the material discussed here.

Historically, the formalism which first arose for the material we discuss is that of measured foliations in surfaces. The reader is referred to [FLP] for an extensive treatment of measured foliations. Roughly, a measured foliation in a surface F is a one-dimensional foliation of F (with singularities of specified types) together with a transverse measure, which assigns to each arc transverse to the foliation a nonnegative real number subject to certain conditions (which are analogous to the conditions on the transverse measure of a measured lamination). The prototypical example of a measured foliation is given by the level sets of a harmonic function on F, where the transverse measure is given by integrating the conjugate differential along arcs in F. Another related prototypical example is given by the horizontal (or vertical) trajectories of a holomorphic quadratic differential on a Riemann surface. (See [St] for an extensive treatment of quadratic differentials, which are deeply connected with the material of this volume.)

A certain equivalence relation (generated by isotopy and Whitehead moves) on measured foliations is considered, and the space of equivalence classes of measured foliations in F is denoted $\mathcal{MF}(F)$. One also considers a corresponding space $\mathcal{MF}_0(F)$ of equivalence classes of measured foliations of compact support. The natural $\mathbb{R}_+$-action on transverse measures leads to corresponding spaces $\mathcal{PF}(F)$ and $\mathcal{PF}_0(F)$ of projective measured foliations and projective measured foliations of compact support.

As the reader has probably guessed, there are identifications

$$\mathcal{ML}(F) \sim \mathcal{MF}(F) \text{ and } \mathcal{ML}_0(F) \sim \mathcal{MF}_0(F),$$
$$\mathcal{PL}(F) \sim \mathcal{PF}(F) \text{ and } \mathcal{PL}_0(F) \sim \mathcal{PF}_0(F).$$

Indeed, given a measured train track (τ, μ) in F, we have seen in this volume how to construct an associated measured geodesic lamination, and we let $(N, \mathcal{F}, \mathcal{F}^\perp)$ be the bifoliated neighborhood of τ in F considered in this construction. Collapsing each region of $F - \tau$ onto a spine as indicated in Figure E, the measured foliation $\mathcal{F}$ gives rise to a measured foliation of F itself (and the equivalence class of this measured foliation in F is well-defined). In fact, the identifications above are induced by these constructions, and we henceforth make these identifications without further comment.

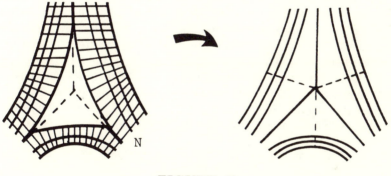

FIGURE E

It is worth emphasizing that one must consider equivalence classes of measured foliations, as above, to define $\mathcal{MF}_0(F)$. In contrast, one may regard measured geodesic laminations as a more intrinsic formulation, since the space $\mathcal{ML}_0(F)$ is exactly the space of measured geodesic laminations (no equivalence relation being necessary). In analogy to the situation with measured foliations, we have considered measured train tracks up to combinatorial equivalence in this volume. At the same time, we have seen that a complete train track in F may be thought of as a chart on the manifold $\mathcal{ML}_0(F)$.

Recall the Teichmüller space $\mathcal{T}(F)$ (see [A], for instance) of hyperbolic structures on (a fixed surface) F modulo push-forward by orientation-preserving diffeomorphisms which are isotopic to the identity. This space is diffeomorphic to an open ball of dimension $6g - 6 + 2s$ if $F = F_g^s$. There are imbeddings

$$\mathcal{ML}_0(F) \to [\mathbb{R}_+ \cup \{0\}]^{\mathcal{S}'(F)}$$
$$(G, \lambda) \mapsto V_{(G,\lambda)}(\cdot)$$

and

$$T(F) \to [\mathbb{R}_+ \cup \{0\}]^{\mathcal{S}'(F)}$$
$$\rho \mapsto \ell_\rho(\cdot),$$

where $\mathcal{S}'(F)$ is the collection of all connected multicurves in F, $\ell_\rho(c)$ denotes the ρ-length of the geodesic in the isotopy class c, and $V_{(G,\lambda)}$ denotes the variation function (as in §2.8) of (G, λ). In fact, these imbeddings have disjoint images, and one uses them to prove that the sphere $\mathcal{PL}_0(F)$ forms the collection of ideal points in a compactification of the open ball $T(F)$ in such a way that

$$\bar{T}(F) = \mathcal{PL}_0(F) \cup T(F)$$

is a closed ball. See [FLP] for a proof of these fundamental results.

The mapping class group $MC(F)$ of orientation-preserving diffeomorphisms of F modulo isotopy acts on $T(F)$ as well as on the various spaces considered above, and a fundamental fact due to Kerckhoff [K1] is that the action of $MC(F)$ on $T(F)$ extends continuously to an action of $MC(F)$ on the closed ball $\bar{T}(F)$. If $f \in MC(F)$, then we let $f_\#$ (and $\bar{f}_\#$, repectively) denote the induced map on $\mathcal{ML}_0(F)$ (and on $\bar{T}(F)$). Since our construction of measured geodesic laminations is natural for the action of $MC(F)$, if (τ, μ) is a measured birecurrent train track, then f induces a bijection f_* from the set of branches of τ to those of $f(\tau)$, and

$$f_\#([\tau, \mu]) = [f(\tau), \mu \circ f_*^{-1}],$$

where $[\tau, \mu] \in \mathcal{ML}_0(F)$ denotes the equivalence class of (τ, μ). It follows that $MC(F)$ acts as a group of symplectomorphisms of $\mathcal{ML}_0(F)$ (for the symplectic structure discussed in §3.2). Furthermore, the "intersection pairing" on $\mathcal{ML}_0(F)$ (discussed briefly in the remark before Lemma 3.4.3) is evidently $MC(F)$-invariant as well.

We mention parenthetically that T also admits several interesting $MC(F)$-invariant geometric structures (cf. [A]). Among these is a Kähler metric, called the "Weil-Petersson metric"; see [Wo], for instance. Provided the surface F has at least one puncture, the Weil-Petersson Kähler two-form extends (in an appropriate sense) to the symplectic structure on $\mathcal{ML}_0(F)$; see [PP2].

The induced action of $MC(F)$ on $\mathcal{ML}_0(F)$ is thus rich with invariant geometric structures, and we next discuss this action. The most naive point of view is to consider the action of $MC(F)$ on one of our many coordinatizations of $\mathcal{ML}_0(F)$. In fact, there are generators of the mapping class group due to Dehn, and we may ask for the representation of these generators as transformations on the Dehn-Thurston coordinates of $\mathcal{ML}_0(F)$. The Addendum sketches the derivation and gives these explicit formulas (which are rather complicated PIL expressions). The material of this paragraph was the subject of [P1].

A more elegant point of view on the $MC(F)$-action on $\mathcal{ML}_0(F)$ involves questions of the dynamics of this action. Insofar as $\bar{T}(F)$ is a closed ball, $\bar{f}_\#$ has a fixed point in $\bar{T}(F)$ by Brouwer's Fixed-Point Theorem for any $f \in MC(F)$. Thus, any $\bar{f}_\#$ must either leave invariant a hyperbolic structure (and must therefore be finite-order) or must leave invariant a projective measured lamination of compact support. If the fixed point of $\bar{f}_\#$ lies "at infinity" (that is, in $\mathcal{PL}_0(F)$), then there are crucial differences depending on whether the underlying lamination has any closed leaves. A point $(G, \lambda) \in \mathcal{ML}_0(F)$ is said to be *arational* if it has no closed leaves.

The fundamental classification of conjugacy classes in $MC(F)$ is

Thurston's Classification of Homeomorphisms: *Given an element of $MC(F)$, there is a representative homeomorphism f satisfying one of the following.*

f is finite-order.

f is reducible, *by which we mean that there is some multicurve in F which is invariant by $f_\#$.*

f is pseudo-Anosov, *by which we mean that there are a pair of transverse arational measured geodesic laminations $(G_\pm, \lambda_\pm)$ and a number $\Lambda > 1$ so that*

$$f_\#(G_\pm, \lambda_\pm) = \Lambda^{\pm 1}(G_\pm, \lambda_\pm).$$

In the reducible case, one can take a sufficiently high power n of f to arrange that f^n fixes each subsurface complementary to a corresponding multiple curve. The composition f^n induces a "component map" on each such subsurface in the natural way, and each component map is either pseudo-Anosov or finite-order. See [FLP] for a complete treatment of this result.

The train track theory is especially well-suited to the analysis of reducible and pseudo-Anosov mapping classes as we next see. Suppose that f is a diffeomorphism of F and that τ is a birecurrent train track in F so that $f(\tau) < \tau$, say with incidence matrix M, as in §2.1. (One can often find such a train track by experimentation.) We can spectrally analyze M in search of eigenvectors (whose corresponding measures satisfy the switch conditions on τ) which correspond to invariant multicurves and, more generally, to (projectively) invariant measured geodesic laminations.

In particular, if the mapping class of f is reducible, say with invariant multicurve c and $c < \tau$, then c must be represented by some integral eigenvector of M with eigenvalue unity.

On the other hand, if the mapping class of f is pseudo-Anosov and $\tau > f(\tau)$, then $G_+ < \tau$ and τ must fill F. (Given a pseudo-Anosov map, [PP1;Theorem 4.1] guarantees the existence of such a train track.) Moreover, it turns out that (there is a choice of supporting map so that) the incidence matrix M is "Perron-Frobenius" (also called "primitive irreducible") in the sense that it has nonnegative entries and some positive power has strictly positive entries. It is classical (see [G], for instance) that such a matrix has a unique positive eigenvector whose corresponding real eigenvalue agrees with the spectral radius of M. Furthermore, its corresponding eigenvector is actually a transverse measure μ on τ which gives rise to the lamination (G_+, λ_+), and its eigenvalue is none other than Λ. Dually, since τ carries $f(\tau)$, it follows from Proposition 3.4.2 that the dual bigon track τ^* is carried by the dual bigon track $[f(\tau)]^*$, and $G_- < \tau^*$, so (G_-, λ_-) gives rise to some tangential measure on τ by Lemma 3.4.3.

Conversely, we briefly discuss some material from [P2], which gives a method of recognizing pseudo-Anosov maps. Let us suppose that f is a homeomorphism of F and τ is a birecurrent train track filling F which splits (with no collisions) to $f(\tau)$ so that the natural incidence matrix M for the carrying $\tau > f(\tau)$ is Perron-Frobenius. The spectral radius Λ of M must be greater than unity since M is an integral matrix, and the eigenvector of M corresponding to Λ gives rise to a positive transverse measure μ on τ. By Proposition 3.4.2, the transpose M^t is an incidence matrix for the carrying $\tau^* < [f(\tau)]^*$ of the dual bigon tracks, so M^t is also an incidence matrix for the carrying $[f^{-1}(\tau)]^* < \tau^*$. Of course, M^t is also Perron-Frobenius, its spectral radius is also Λ, and the eigenvector corresponding to Λ gives rise to a positive tangential measure ν on τ.

Build the bifoliated neighborhood $(N, \mathcal{F}, \mathcal{F}^\perp)$ as in our construction of measured geodesic laminations, but take the rectangle corresponding to the branch b of τ of length $\nu(b)$ (and width $\mu(b)$ as before) for each branch b. It turns out that one can collapse complementary regions as in Figure E in such a way that the ν-lengths of identified arcs agree, so $\mathcal{F}$ and $\mathcal{F}^\perp$ give rise to corresponding transverse measured foliations (of the same name) on F itself. (One must use the dynamics of the action of f on N to see that there exists such a collapse; see the remark following Lemma 3.4.3.) It follows that $\mathcal{F}$ and $\mathcal{F}^\perp$ are arational, and we let (G_+, λ_+) and (G_-, λ_-) denote the corresponding arational measured geodesic laminations. By construction,

$$f_\#(G_\pm, \lambda_\pm) = \Lambda^{\pm 1}(G_\pm, \lambda_\pm),$$

so f is pseudo-Anosov. We remark parenthetically that one can probably relax the condition above that τ splits to $f(\tau)$ to allow shifting as well and still conclude that f is pseudo-Anosov.

This gives yet another aspect of the duality between transverse and tangential measure and further explains our remark (at the end of §3.4) that tangential measures may be interpreted as the lengths of rectangles in the bifoliated neighborhood.

We turn finally to deformation theory. There are certain deformations of complex structures on a Riemann surface called "earthquakes" which correspond to sliding complementary regions of a measured geodesic lamination along the leaves of the lamination; see [K3] for a precise definition and further information. Earthquakes are thus a natural generalization of the classical Fenchel-Nielsen deformations (see [A], for instance). Earthquake deformations are parametrized by the space $\mathcal{ML}_0(F)$ itself, and they fill out the tangent space at a point of $\mathcal{T}(F)$; see [K3]. (The cotangent space is exactly the space of holomorphic quadratic differentials mentioned above.) The fundamental fact also due to Kerckhoff [K2] is that the hyperbolic length function $\ell.(c)$ is convex along earthquake paths, for each multicurve c in F.

Furthermore, Papadopoulos has shown [Pa] that Fenchel-Nielsen flows on $\mathcal{T}(F)$ extend continuously to flows on $\bar{\mathcal{T}}(F)$, and these flows are hamiltonian for the symplectic structure of $\mathcal{ML}_0(F)$.

ADDENDUM
THE ACTION OF MAPPING CLASSES ON $\mathcal{ML}_0$

In this section, we give the explicit formulas from [P1] for the action of the mapping class group on our basic parameters; see [P3] for a survey of this. A primary motivation is that essentially all of the required material has been developed here, and our intent is to include certain basic formulas.

Suppose that K_i is a pants curve in the basis $\mathcal{A}$ for $\mathcal{S}(F)$, and consider a "(right) Dehn twist" T_i (see, for instance, [L]) along K_i. As observed by Dehn [D], the action of T_i on Dehn-Thurston coordinates $\{m_j, t_j\}_1^N$ on $\mathcal{S}(F)$ is linear

$$T_i^{\pm 1} : \; t_i \mapsto t_i \pm m_i,$$

the other coordinates being unchanged.

Consider the two changes of basis for multiarcs (not just multicurves) from the basis $\mathcal{A}$ to the basis $\mathcal{A}'$ indicated in Figure A on the surfaces $F_1^{0,1}$ and $F_0^{0,4}$.

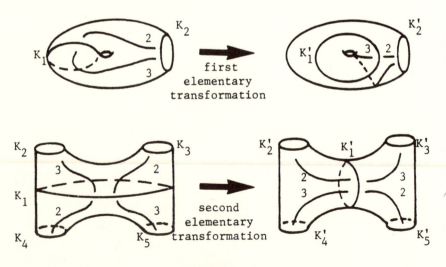

FIGURE A

210

These changes of basis are called the *first* and *second elementary transformations*, as indicated. It is an easy exercise (see [P1]) to produce a family $\{\mathcal{A}_k\}_1^K$ of bases differing only by compositions of elementary transformations so that for each Dehn twist in a generating family (see [H], for instance, for a family of Dehn twists whose associated mapping classes generate $MC(F)$), the corresponding curve occurs as a pants curve in $\mathcal{A}_k$, for some $k = 1, \ldots, K$. To compute the action of $MC(F)$ on Dehn-Thurston coordinates with respect to some basis among $\{\mathcal{A}_k\}$, it therefore suffices to compute the effect of each of the elementary transformations on Dehn-Thurston coordinates.

There is the following hybrid of our coordinatizations of $\mathcal{ML}_0$: record a Dehn-Thurston twisting number as before for each pants curve and record the basic parameters on each complementary pair of pants. For each of the pants curves K_i in $\mathcal{A}$ and K_i' in $\mathcal{A}'$, let t_i and t_i', respectively, denote these twisting numbers. For the first elementary transformation, let λ_{mn} and λ'_{mn} denote the basic parameters in the pair of pants of the basis $\mathcal{A}$ and $\mathcal{A}'$, respectively. For the second elementary transformation, let

κ_{mn} be the basic parameters in the top pants for the basis $\mathcal{A}$,

λ_{mn} be the basic parameters in the bottom pants for the basis $\mathcal{A}$,

κ'_{mn} be the basic parameters in the right pants for the basis $\mathcal{A}'$,

λ'_{mn} be the basic parameters in the left pants for the basis $\mathcal{A}'$,

let $\wedge$, $\vee$, respectively denote the binary infimum and supremum operations, and let $\text{sgn}(x)$ denote the sign of $x \in \mathbb{R}$ (where $\text{sgn}(0)$ depends on conventions made below).

We can finally state the main results of [P1].

Theorem A.1: *The first elementary transformation is given by the following formulas.*

$$\lambda'_{11} = \{r - |t_1|\} \vee 0$$
$$\lambda'_{12} = \lambda'_{13} = L + \lambda_{11} - \lambda'_{11}$$
$$\lambda'_{23} = |t_1| + \lambda'_{11} - L$$
$$t'_2 = t_2 + \lambda_{11} + \{(L - \lambda'_{11}) \wedge t_1\} \vee 0$$
$$t'_1 = -\text{sgn}(t_1) \{\lambda_{23} + L - \lambda'_{11}\},$$

where $L = \lambda_{12} = \lambda_{13}$ and $\text{sgn}(0)$ is taken to be -1.

Actually, Theorem A.1 is rather easy. It is our next result which requires massive computations.

Theorem A.2: *The second elementary transformation is given by the following formulas.*

$$\kappa'_{11} = \kappa_{22} + \lambda_{33} + \{L - \kappa_{13}\} \vee 0 + \{-L - \lambda_{12}\} \vee 0$$

$$\kappa'_{22} = \{L \wedge \lambda_{11} \wedge (\kappa_{13} - \lambda_{12} - L)\} \vee 0$$

$$\kappa'_{33} = \{-L \wedge \kappa_{11} \wedge (\lambda_{12} - \kappa_{13} + L)\} \vee 0$$

$$\kappa'_{23} = \{\kappa_{13} \wedge \lambda_{12} \wedge (\kappa_{13} - L) \wedge (\lambda_{12} + L)\} \vee 0$$

$$\kappa'_{12} = -2\kappa'_{22} - \kappa'_{23} + \kappa_{13} + \kappa_{23} + 2\kappa_{33}$$

$$\kappa'_{13} = -2\kappa'_{33} - \kappa'_{23} + \lambda_{12} + \lambda_{23} + 2\lambda_{22}$$

$$\lambda'_{11} = \lambda_{22} + \kappa_{33} + \{K - \lambda_{13}\} \vee 0 + \{-K - \kappa_{12}\} \vee 0$$

$$\lambda'_{22} = \{K \wedge \kappa_{11} \wedge (\lambda_{13} - \kappa_{12} - K)\} \vee 0$$

$$\lambda'_{33} = \{-K \wedge \lambda_{11} \wedge (\kappa_{12} - \lambda_{13} + K)\} \vee 0$$

$$\lambda'_{23} = \{\lambda_{13} \wedge \kappa_{12} \wedge (\lambda_{13} - K) \wedge (\kappa_{12} + K)\} \vee 0$$

$$\lambda'_{12} = -2\lambda'_{22} - \lambda'_{23} + \lambda_{13} + \lambda_{23} + 2\lambda_{33}$$

$$\lambda'_{13} = -2\lambda'_{33} - \lambda'_{23} + \kappa_{12} + \kappa_{23} + 2\kappa_{22}$$

$$t'_2 = t_2 + \lambda_{33} + \{(\lambda_{13} - \lambda'_{23} - 2\lambda'_{22}) \wedge (K + \lambda'_{33} - \lambda'_{22})\} \vee 0$$

$$t'_3 = t_3 - \kappa'_{33} + \{(L + \kappa'_{33} - \kappa'_{22}) \vee (\kappa'_{23} + 2\kappa'_{33} - \lambda_{12})\} \wedge 0$$

$$t'_4 = t_4 - \lambda'_{33} + \{(K + \lambda'_{33} - \lambda'_{22}) \vee (\lambda'_{23} + 2\lambda'_{33} - \kappa_{12})\} \wedge 0$$

$$t'_5 = t_5 + \kappa_{33} + \{(\kappa_{13} - \kappa'_{23} - 2\kappa'_{22}) \wedge (L + \kappa'_{33} - \kappa'_{22})\} \vee 0$$

$$t'_1 = \kappa_{22} + \lambda_{22} + \kappa_{33} + \lambda_{33} - \{\lambda'_{11} + \kappa'_{11} + (t'_2 - t_2) + (t'_5 - t_5)\}$$
$$\quad + \left[\mathrm{sgn}(L + K + \lambda'_{33} - \lambda'_{22} + \kappa'_{33} - \kappa'_{22})\right](t_1 + \lambda'_{33} + \kappa'_{33}),$$

where $L = \lambda_{11} + t_1$, $K = \kappa_{11} + t_1$ *and*

$$\mathrm{sgn}(0) = \begin{cases} +1, & \text{if } \lambda_{12} - 2\kappa'_{33} - \kappa'_{23} \neq 0; \\ -1, & \text{otherwise.} \end{cases}$$

Remark: The computations in [P1] are done in the setting of multicurves/multiarcs and involve explicit homotopies. The formulas apply more generally to measured geodesic laminations and train tracks from the density of multicurves (Theorem 3.1.3) and multiarcs (cf. §1.8). (We take this opportunity to point out that there is a gap in the proof of [P1;Lemma 4.1]; this gap is much more than filled by [FHS].) A train track treatment of the second elementary transformation (using technical facts from [P1;§7]) might well be simpler than our approach with curves and arcs. However, the formulas for the elementary transformations will not simplify very much (if at all).

One can easily derive PIL expressions for the action of $MC(F)$ on our hybrid coordinates using the previous theorems. Indeed, each generating Dehn twist is given as a conjugate $f \circ T \circ f^{-1}$, where f is a (known) composition of the elementary transformations (which are PIL), and T is a linear map (associated to a Dehn twist on a pants curve as above). It is straight-forward to implement this solution on the computer, and many trends predicted by Thurston's Conjugacy Classification of mapping classes (see the Epilogue) are exhibited by experimenting with this code, so it is instructive to play with. It might also be interesting to try symbolic manipulation techniques in the hopes of simplifications for the PIL conjugates above.

BIBLIOGRAPHY

[A] W. Abikoff, "The Real-Analytic Theory of Teichmüller Space", Lecture Notes in Mathematics # 820, Springer-Verlag, Berlin (1980).

[Be] A. Beardon, "The Geometry of Discrete Groups", Graduate Texts in Mathematics # 91, Springer-Verlag, New York (1983).

[Bo] F. Bonahon, *Bouts de variétés hyperboliques de dimension 3*, Annals of Mathematics **124** (1986), 71-158.

[C] A. Casson, "Automorphisms of Surfaces after Nielsen and Thurston", Cambridge Univ. Press, Cambridge (1988).

[Cb] A. Casson, *Automorphisms of Surfaces after Nielsen and Thurston*, Lecture notes from Austin, Texas; notes by S. Bleiler (1983).

[D] M. Dehn, Lecture Notes from Breslau, 1922. The Archives of The University of Texas at Austin.

[E] D. B. A. Epstein, *Curves on two-manifolds and isotopies*, Acta Mathematia **115** (1966), 83-107.

[FHS] M. Freedman, J. Hass, P. Scott, *Closed geodesics on surfaces*, Bulletin of the London Mathematical Society **14** (1982), 385-391.

[FLP] A. Fathi, F. Laudenbach, V. Poenaru, et al., "Travaux de Thurston sur les Surfaces", Asterisque **66-67**, Societe Mathematique de France, Paris (1979).

[G] V. F. Gantmakher, "Theory of Matrices", volume II, Chelsea, New York (1959).

[GP] V. Guillemin and A. Pollack, "Differential Topology", Prentice-Hall,

New Jersey (1974).

[HT] A. Hatcher and W. P. Thurston, *A presentation for the mapping class group of a closed orientable surface*, Topology **19** (1980), 221-237.

[H] S. Humphreys, *Generators for the mapping class group*, in the volume "Topology of Low-Dimensional Manifolds", Lecture Notes in Mathematics #722, Springer-Verlag, Berlin (1979), 44-47.

[K1] S. Kerckhoff, *The asymptotic geometry of Teichmüller space*, Topology **19** (1980), 23-41.

[K2] ——, *The Nielsen realization problem*, Annals of Mathematics **117** (1983), 235-265.

[K3] ——, *Earthquakes are analytic*, Commentarii Mathematici Helvetici **60** (1985), 17-30.

[L] W. B. R. Lickorish, Proceedings of the Cambridge Philosophical Society *A finite set of generators for the homeotopy group of a 2-manifold* **60** (1964), 769-778; *Corrigendum* **62** (1966), 679-781.

[M] H. Masur, *Interval exchange transformations and measured foliations*, Annals of Mathematics **115** (1982), 169-200.

[N] J. Nielsen, *Surface transformation classes of algebraically finite type*, Danske Vid. Selsk. Mat.-Fys. Medd. **XXI** (1944), 1-89.

[Pa] A. Papadopoulos, *Geometric intersection functions and hamiltonian flows on the space of measured foliations on a surface*, Pacific Journal of Mathematics **124** (1986), 375-402.

[PP1] A. Papadopoulos and R. C. Penner, *A characterization of pseudo-Anosov foliations*, Pacific Journal of Mathematics **130** (1987), 359-377.

[PP2] ——, *La forme symplectique de Weil-Petersson et le bord de Thurston de l'espace de Teichmüller*, Comptes Rendus, Acad. Sci. Paris **312**, serie 1 (1991), 871-874.

[P1] R. C. Penner, "A Computation of the Action of the Mapping Class Group on Isotopy Classes of Curves and Arcs in Surfaces", thesis, Massachusetts Institute of Technology (1982).

[P2] ——, *A construction of pseudo-Anosov homeomorphisms*, Transactions of the American Mathematical Society **310** (1988), 179-197.

[P3] ——, *The action of the mapping class group on curves in surfaces*, L'Enseignement Mathematique **30** (1984), 39-55.

[Sp] E. Spanier, "Algebraic Topology", McGraw-Hill, New York (1966).

[St] K. Strebel, "Quadratic Differentials", Springer-Verlag, Berlin (1984).

[T1] W. P. Thurston, *On the geometry and dynamics of diffeomorphisms of surfaces*, Bulletin of the American Mathematical Society **19** (1988), 417-431.

[T2] ——, "Three-dimensional Geometry and Topology", Annals of Mathematical Studies, Princeton Univ. Press (to appear).

[Tg] ——, Lecture notes from Boulder, Colorado; notes by W. M. Goldman (1981).

[Wo] S. Wolpert, *On the symplectic geometry of deformations of a hyperbolic surface*, Annals of Mathematics **117** (1983), 207-234.